Lighting Modern Buildings

Lighting Modern Buildings

Derek Phillips

Architectural Press

OXFORD   AUCKLAND   BOSTON   JOHANNESBURG   MELBOURNE   NEW DELHI

Architectural Press
An imprint of Butterworth-Heinemann
Linacre House, Jordan Hill, Oxford OX2 8DP
225 Wildwood Avenue, Woburn, MA 01801–2041
A division of Reed Educational and Professional Publishing Ltd

 A member of the Reed Elsevier plc group

First published 2000

British Library Cataloguing in Publication Data
Phillips, Derek
   Lighting modern buildings
   1. Interior lighting
   I. Title
   729.2'8

ISBN 0 7506 4082 0

Library of Congress Cataloguing in Publication Data
A catalogue record for this book is available from the Library of Congress

Composition by Scribe Design, Gillingham, Kent
Printed and bound in Great Britain

Contents

Foreword

Lighting has been one of the main modes of expression of the architect over the centuries. The character of interior spaces can be enhanced by controlling the admission of daylight by means of the form of a building and the size, position and aspect of openings in its fabric. The range of expression is great, from the evocative lighting of the Gothic cathedral and the drama of the Baroque church to the simple quietude of a domestic interior. The lit effect, the interplay of light and shadow, is a response to functional and emotional needs derived from a unified design approach.

In previous centuries lighting after dark was provided by incandescent sources, torches, candles, oil lamps and latterly by gas. By today's standards opportunities for expression were limited. The advances that have occurred in lighting technology during the last hundred years have had a great influence on the way we live. Electric light is readily available at the touch of a switch; it can be varied in intensity and, with the right equipment, it can be redirected, refocused and its colour changed, or it can be piped remotely from its source and redistributed whilst its duration can be controlled. The efficiency of lamps has shown constant improvement and the range of types has been developed to satisfy growing and changing needs. As this book shows, the palette for designing with light is extensive.

Mounting concern with global warming and pollution have placed emphasis upon energy conservation and have brought about a serious reconsideration of the relationship between daylight and electric light. Lighting design has been confirmed as an essential part of the overall process of building design. The architect today needs an understanding not only of lighting as an art form, but also of its environmental implications, its technology and its hardware.

This book responds to these needs. It should appeal not only to professional designers but to all those with inquiring minds and to those who are sensitive to their visual surroundings. Key issues which influence the development of interior building design are examined and the author explains a lighting design strategy based both on the physical aspects of seeing and on perception, the interpretation by the mind and the emotional response. He discusses the relationship between natural and artificial lighting and their integration with the building fabric, structure and other services. The all-important interactive nature of the design process is emphasized.

As an architect/lighting designer and the creator of many notable schemes, both in this country and overseas, the author has drawn upon his experience in selecting, illustrating and describing significant examples of work by many designers for critical study. These are presented not as copy-book exercises, but to stimulate discussion in each case about the approach, principles and practicalities which fashioned the final result.

Words and pictures together make this a fascinating and informative book for all with an interest in creative design; it is a worthy successor to the author's *Lighting in Architectural Design* which assessed the state of the art some thirty-five years ago. As will be seen, much has happened since then.

James Bell
Emeritus Professor of Architecture
The University of Manchester

Preface

My earlier book, *Lighting in Architectural Design* published by McGraw Hill in 1964, resulted from a paper on architectural education given to the RIBA and the IES in London.[1] The paper made clear that due to the disparate training of architects and lighting engineers, there was a lack of understanding between the two professions, resulting in a lack of co-operation and ultimately in poor lighting design.

A lot has changed in 35 years: some improvement has been made in the training of architects in this field, although more needs to be done; whilst a new profession of 'Independent Lighting Designer' (IALD)[2] has emerged, leaving the original profession of 'Lighting Engineer' (generally associated with the manufacturers and utilities) to cope with the essential development of light sources, the design of lighting equipment and the distribution and control of light . . . the 'building services' side. Indeed, in the UK this is now dealt with by the Chartered Institute of Building Services Engineers (CIBSE), the Illuminating Engineering Society (IES) having been merged with the Institute of Heating and Ventilation Engineers to form the new body.

A useful future development would be in the formation of a single body to look after the interests of both daylight and artificial light, and to encompass all those architects, engineers and lighting designers who have a functional role to play in the design of buildings.

My earlier book dealt in some depth with methods of calculation for both daylight and artificial light, but this is now thought to be unneccessary, since computer generated techniques are widely available. However, computer aided design is no substitute for the thought which architects and designers must give to the lighting concept for a building. At best it can be used to present an impression of a lit space, or exterior, to a client, to explain the development of the lighting concept, whilst providing a necessary check on preliminary estimates of light values.

In 1964 one might have been forgiven for believing that the future development of electric light sources would diminish the influence of daylight, to the point where artificial forms of light might take over the role of lighting buildings both during the day as well as at night, leaving the role of daylight more concerned with the view out of a building. This approach is now seen to be far from the case. It is not only due to the

---

[1] Lighting in Buildings. Training and Practice Trans. IES (GB), Vol 21 No 3 1956.
[2] International Association of Lighting Designers (IALD).

world scarcity of energy and the need to reduce carbon emissions, that there is an imperative to use the natural resource of daylight more efficiently; there is an overwhelming case to be made now, as in the past, for the use of daylight in our buildings for environmental reasons unconnected with the price of fuel.

Part two of the book is devoted to the lighting design of a number of architectural projects, from those of the early modern movement to todays' hi-tech solutions. The relationship between lighting design during the day and at night is explored, to illustrate the theories of integration between the two developed in the earlier part of the book.

Written at the end of a career as an architect and lighting designer (from 1952 to 1993), the book draws on experience gained while living through a period of intense lighting development, leading from the original paper in 1956 up to the millennium. Its purpose will have been served if it leads to a greater understanding of the critical importance of all forms of lighting in architectural design.

Acknowledgements

On the completion of a work of this nature, where help has been received from so many quarters, it is impossible to recognize the many architects and lighting designers individually who have contributed to the book.

The Case Studies must speak for themselves, and the architects and lighting designers responsible are acknowledged, but particular thanks must go to those practices who have co-operated so generously in providing the photographs and drawings which illustrate these projects. The most difficult part of writing a book where illustrations play so vital a part, is persuading designers to spend the necessary time in seeking out schemes which they may have completed many years before.

Having said this, there are several people who have been generous with their time, in reading and correcting the early theoretical chapters. I must mention Prof James Bell who helped with the chapter on Daylighting; Tony Willoughby and John Howard who have eliminated many of the mistakes I would otherwise have made on artificial lighting; and finally David Loe, and Kit Cuttle from Rensselaer who have helped to clarify my thoughts on the analysis of the 'architectural problem', and my approach to 'seeing and perception'.

Many thanks to all those who have helped in one way or another to get this book printed.

Part 1

1 Introduction

When *Lighting in Architectural Design* was originally published in 1964 there was a need to emphasize the fundamental importance of the architect, as it was considered that his role had been usurped by the increasingly influential lighting engineer who was supported by the tremendous advances in the field of electric light sources. The architect had lost the initiative.

The question was posed as to the outcome of an unlimited supply of artificial light. Artificial light would become the primary light source for vision in all architectural programmes demanding efficient seeing both during the day as well as after dark, leaving natural light to maintain preservation of spatial unity, modelling and the psychological advantages associated with contact with the natural environment outside, the view out.

This basic assumption has been proved false. Indeed, the opposite may be closer to the truth as natural light is, if anything, becoming more important because of the crisis in world energy. Far from losing its importance with advances in artificial lighting, as might have been perceived in 1964, it has gained in influence due to a renewed understanding of its true values. The starting point for any lighting design today is, as throughout history, the 'natural' source.

An understanding of lighting demands a sense of history, and the author's book *Lighting Historic Buildings* (Phillips, Derek, *Lighting Historic Buildings*, Architectural Press, 1997) traces the development of lighting in buildings from earliest times to the end of the nineteenth century. This current book covers the twentieth century, leading up to the millennium. It looks at the early development of electric light sources and how they have been used in buildings, initially as separate systems and later in association with daylight in integrated solutions.

Lighting design is fundamental to the success of any building. This hardly needs stressing since the way the interior of a building is perceived depends upon the way in which it is lit. Light is as much a building material as the structure of which it is made, since light, when reflected from a structure's surfaces and edges, provides the information upon which we act. It is light that can make a building bright and airy or dull and gloomy; it is light that enables us to perform, and without it the building would cease to function.

This book starts with an exploration of the basic human needs of vision and the perception of our exterior world – the emotional and intellectual together with the physical requirements of vision – since lighting must satisfy our needs at both levels. The engineering approach which may suggest that human needs will be satisfied if the needs of vision as evidenced by the level of light on the task, are met, falls far short of what is required. Here the physical factors of 'seeing' are distinguished from the less tangible factors of 'perception' which the architect must consider.

Whilst an understanding of the role of daylight is fundamental, a knowledge of the various forms and properties of artificial light not only at night but during the day is essential. In early buildings there was one form of light – daylight – during the day and another – artificial light – at night. No attempt was made to integrate the two. Both had their own essential but separate qualities. The situation today is very different for many reasons, not least because of the possibilities posed by modern structure. It is now impossible to consider the daytime lighting of the majority of buildings without an integrated solution.

It is necessary to explore the nature of artificial lighting, its own essential qualities and its relationship with building structure and servicing to understand how the day and night lighting must be designed to provide optimum solutions.

The interior of a building at night does not necessarily have to imitate its daylight appearance. In 'work places' it may be thought that conditions should be strictly controlled as far as possible, night and day, to create a standard environment, but this is not the case in the majority of buildings, where man's emotional needs suggest that a change of appearance after dark may be beneficial.

Artificial light is in a state of constant development and the effects now available will no doubt be as out of date in 2050 as those of 1950 are to us today. For this reason, whilst aspects of artificial lighting – such as available light sources, methods of lighting and controls – are discussed, it is more in terms of principle than in detail; change must be accepted.

The important disciplines associated with structure and its relationship with installation and maintenance are emphasized, not forgetting other building services, such as ventilation and acoustics – the trilogy of 'Heat, Light and Sound.'

A major portion of the book is devoted to 'design'. Not wishing to attempt to create 'pattern books' in the eighteenth century sense, this is achieved by analysis of a large number of well documented examples of building solutions, where the daylighting and artificial lighting has been considered separately and together and an integrated design achieved. Sufficient technical detail is provided to permit an understanding of the design principles of each scheme.

The schemes illustrated set out to cover the development of natural and artificial lighting in buildings throughout the twentieth century, from those of the early modern movement in the 1930s to the latest 'Hi-tech' developments. They vary from small domestic buildings, churches and workplaces to those devoted to leisure and sport; the aim is to develop the design concepts set out in the earlier chapters, and to lead the way forward for architects in the future.

2 Analysis

Looking back at the earliest days of lighting design, those features that we now recognize as complex were evident in the design solutions of the time, and the daylighting design for an Egyptian Temple or Sta. Sophia in Constantinople in the sixth century would have been thought of as state of the art then. What has changed today more than anything is in the priority that must now be attached to the different elements of design.

When monks were illustrating the gospels in the twelfth century they would work by daylight during the day and by candlelight at night. There was insufficient knowledge of the physiological effects of the strain this would have placed on their eyes, but the problems of perception would have been evident in the difficult circumstances in which they worked.

As with all aspects of architecture, 'lighting' may be analysed and its relative importance changed depending upon time, the building's function and its location. The needs of architectural programmes change, as do the details of lighting design; what does not change is the 'unity' that should inform all the elements or the 'whole' of the architect's design.[1] It is this unity or 'whole' that will be experienced by those using a building, each according to their own age and past experience, and at different levels of consciousness.

In setting out an analysis for lighting design it is not possible to provide a linear diagram with a neat start and finish since lighting design is bound up with the architect's circular method of working; there is no finite starting point, although inevitably there must be a finish, otherwise the building would never get built.

The analysis suggested is set out in a circular diagram (Fig 2.1) for whilst each aspect needs to be investigated at some stage, the order in which this may be done will vary with the project, and the architects' own priorities. For this reason, the different elements are not in any numbered sequence, but are identified by the chapter in which the subject is covered later in the book.

SEEING–PERCEPTION

The book makes the important distinction between the physical act of seeing and the more emotional and intellectual act of perception. Seeing is concerned with the biological aspects of the eye, perception with the

---

[1] *Whole*: The philosophy of life of 'Holism' as expounded by Jan Smuts.

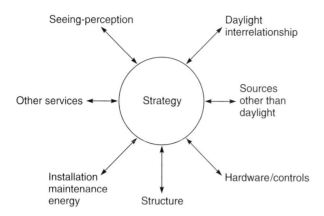

Figure 2.1
Lighting strategy

less tangible elements that make up the whole process or experience of our environment.

Starting with seeing, the most obvious aspect is that there should be enough light to facilitate the performance of a task. This will vary from providing light for the most difficult task, such as that performed in a surgeon's operating theatre, to the most simple, such as a person sitting listening to music at home. Thus, with the level of light, the illumination required will vary from very high to negligible.

Then there are the factors which will affect a person's vision of an object, such as glare which will impair both vision and communication. Glare from both daylight and artificial light must be understood and avoided. The distinction is made between 'glare' which causes disability and glare which causes discomfort.

Colour is important, as is contrast. It is possible to reduce the information gained from a scene by reducing the contrast between what has to be done (the task) and its background. This applies as much to too much light as to too little.

The amount of light a person needs will increase with age and this should be taken into account. But it is not always the amount of light that is important, but often other factors such as contrast, already identified, and clarity. Seeing and health, as well as age, affect the manner of vision.

The element of seeing–perception must be applied to all projects. Although the significance it will have on different problems will vary with the type of architectural programme, it can never be ignored.

DAYLIGHT

The nature of daylight changes. The parameters of daylight are well-known and vary within a given framework, whilst the nature of those sources of light other than daylight are in constant development, together with the hardware associated with them.

The relationship of daylight to the structure of a building and the unique environmental advantages of its use are discussed. Some historical examples are given of the delight that can be derived from the changing daylight patterns both inside and outside the building, associated with the dynamism of the view out.

It is not possible, and never has been, except in engineering terms, to consider the lighting design for a building without assessing the daylight and its relationship with the interior, its modelling and colour. To do this demands an understanding of site orientation, exterior obstruction, climate, the adverse effects of 'glare' and the means to cope with it.

Window design is at the heart of daylighting and various window types are discussed, together with the nature of the new types of glazing now available.

Some guidance is given on the nature of daylight calculation, with particular reference to 'model studies'. These are physical models, because although computer models have been developed for assessing the visual effect of daylight from the sky and sunlight, they are no substitute for seeing the building in three dimensions and assessing it under real sky conditions.

The relationship of daylight to the needs of artificial lighting during the day must be considered, not only in terms of the visual quality of their relationship – the nature of the hardware – but also in terms of their combined energy use. These matters are dealt with elsewhere, but they are related to the daylight strategy to be adopted

SOURCES OTHER THAN DAYLIGHT (ARTIFICIAL LIGHT)

Just as daylight is a critical design issue, so too are artificial sources. Indeed, Professor Hardy was quoted as saying in the 1970s that the first decision an architect has to make is what type of light source should be adopted and what level of illumination is required. He may have been thinking of the artificial light source and level of lighting required at night since any decisions made would affect other decisions on heating and ventilation. Any decisions on structure, installation and maintenance might follow.

In view of today's preoccupation with daylight it is certain that the initial decision an architect has to make relates to daylight and its inter-relationship with the internal artificial lighting of the building. It is essential, therefore, that the artificial light sources now available are identified, together with their characteristics, to enable an architect to make a sound choice.

Light sources are constantly being developed and the one thing that is certain is that information that may be up-to-date on publication will soon be superseded. This makes it no less necessary to cover what exists now, if only to give an architect a standard by which new lamp development may be judged. Crucial aspects of lamp choice are size, lamp life, efficiency, or energy use, and colour; other points to consider are the capacity to control the lamp and the important element of cost.

Initial lamp cost has to be assessed in terms of how much it costs to use the lamp and the life it has. A simple example is to compare the cost of a GLS incandescent lamp with that of a compact fluorescent (CFL). The CFL will cost perhaps ten times more than the GLS, but because of its longer life and greater efficiency the cost over its life to provide an equivalent illumination level will be lower. Whilst in commercial buildings such analysis (and the resulting long-term economic assessments) will be made, in the home, where there could be considerable energy savings, the initial cost of the CFL lamp continues to remain a stumbling block. Other factors that affect lamp choice are its colour rendering and stability, together with its capacity to be controlled. All new lamps will need to be assessed in this way, and it is by no means the initial cost of the lamp that will be found to be the most important element in its choice.

HARDWARE

Of equal importance to the choice of lamp is the type of hardware associated with it. This must first be analysed in terms of how the lighting equipment is attached to the building structure, whether it is direct mounted, suspended, concealed or portable, decisions which are crucial to the architect's perception of the space he wishes to create. Secondly it is necessary to examine the optical distribution of light from the light fittings.

Decisions on the hardware affect the way the spaces are perceived. It is not only the hardware that is important, although this can be a major design factor for good or bad; it is the distribution of light from the hardware which will determine the information of the space, its appearance, and the way it is perceived.

Associated with the hardware are the 'control systems' now available, which range from the simple on/off switch to the highly sophisticated computer controls which permit light sources to be dimmed according to the amount of daylight available outside the building.

The type of hardware chosen will determine the appearance of the space, making it light and airy, or dull and heavy, comfortable, bland, gloomy or dramatic. Whilst other elements of the interior design will contribute to our perception of the building's interior, it is the nature of the light falling upon the surfaces and edges of the space that will have the greatest influence. The hardware is therefore critical.

An analysis of the methods of lighting and whether light is received from downlights or uplights, wallwashers or spotlights, concealed lighting or local lights, will help to determine the character and quality of the space. There are certain obvious rules that should be followed when lighting a building, but it is in the field of quality that the architect's experience has the greatest contribution to make.

BUILDING STRUCTURE

Perhaps the most important element of all in lighting is the structure of the building, for there is an interrelationship between the method of lighting and the structural form, the former revealing the latter. It is this more than any other aspect which will contribute to the way in which the building is perceived, whether gloomy and depressing or light and airy.

Early structures derived from a daylight strategy and it was not until the middle of the twentieth century that any thought was given to the integration of the means of artificial lighting with structure. Therefore early buildings, from churches to factories, were designed for the admission of natural light through windows or roof lights. Artificial light was at best an addition to the structure, for example in the form of pendant chandeliers, despite the fact that many of these ways of lighting were most beautiful.

However by the 1950s, due to the intense technological development in artificial light sources, both in terms of their efficiency and in the design of associated lighting equipment, it was necessary to take the integration of the artificial light with daylit structure seriously, getting to the point where by the 1960s doubts were being raised as to the use of daylight for functional light in the workplace, and the realization that structures might be designed partially or wholly to exclude natural light. This trend towards the exclusion of natural light was short-lived and it was not long before the architect's natural reluctance to eliminate the joys of daylight, together with the growing need for energy conservation, ensured that both daylight and artificial light were combined to provide a high quality environment.

Internally, there are two main types of lighting structure: the expressed structure which makes a virtue of the real structural elements; and the concealed structure where suspended ceilings and other falsework are used to hide the pipes and ducts of modern servicing. A further aspect of the concealed structure is where the whole structure itself becomes the light fitting and ceilings are used as giant reflectors, using daylight or the artificial source, or both together. It is suggested that a different approach is adopted for each of these two building methods.

It is important for the architect to recognize the way in which light can modify structure, where a wall can be made to have mass and solidity, or be broken up by light. The architect must use his powers of observation and experience if he is to understand the consequences of light distribution and the integrity of structure.

The important impression a building takes on after dark, or the 'second aspect of architecture' must never be forgotten. Modern buildings very often have large elements of glass on their façades, leading to internally-lit glass boxes, and the night effect of a building must be thought of at the planning stage, when small modifications to the structure can often be made to ensure a coherent view.

INSTALLATION AND MAINTENANCE

It is important to acknowledge how the lighting is to be installed and, associated with this and of particular relevance to the long term success of the project, how it will be maintained. Whilst lighting design in the nineteenth century may not have had to take much account of this or indeed the needs of other servicing elements in the building, for example those of heating and ventilation or acoustics, the importance of energy considerations and 'costs in use' make an optimum solution essential today.

Installation covers the problems of electrical distribution and those of the support and location of lighting equipment, problems which are visually associated with the structural integrity or 'unity' of the project. Maintenance is closely related to the method of installation, because unless the lighting is fitted so that it can be maintained, cleaned and serviced at suitable intervals it will not operate properly in the manner designed.

The relay of power to the lamps, the electrical distribution, is at the heart of an installation, and as such must be addressed at the initial design stage. Different methods of electrical distribution are now available, and it is important that where a degree of flexibility is required, as may be the case in a work situation, it should be possible to alter the functional layout of the space without extensive changes to the electrical installation during the life of the building.

Very sophisticated ways of controlling lighting are now available and the appropriate degree of control for the needs of the building must be decided at the installation design stage.

Installation must be considered with the maintenance of the lighting system in mind, since unless the needs of maintenance are addressed at the design stage, they will be difficult or even impossible to solve.

Account needs to be taken of the life of chosen lamps and equipment and of the environment they will be used in. For example, if incandescent lamps are used with a stated life of 1000 hours they must be changed regularly or the lamps underrun to increase their life, with a consequent loss of light and an increase in the maintenance costs.

It is not possible to discuss the lighting installation without looking at costs and energy management. Whilst costs could well be ascribed to other sections they are mentioned here as the question of energy is so bound up with costs, and both can be considered an inherent part of the design decisions regarding the installation. Costs have to be divided into those associated with the initial outlay for the lamps and equipment, the design of the installation, and the installation itself (the capital costs) and the running costs associated with the amount of energy used and the maintenance of the installation. Both types of costs need to be considered, but it should be borne in mind that the capital costs are static and the running costs will continue to rise.

When installing lighting the architect should take an active interest in the important subject of safety, both whilst the building is under construction and when built. Two of the most important dangers are electric shock, and fire, which raises the questions of the means of escape or emergency lighting.

BUILDING SERVICES

Finally there is the relationship of lighting design with other building services, the most important of these perhaps being heating, ventilation and acoustics.

Daylight and artificial lighting design must work in harmony with the needs of the heating and ventilation engineer, not only in terms of the use of energy, but in the co-ordination and location of lighting equipment with heating and ventilation ducting and openings.

Acoustics is also an area where a careful relationship needs to exist with the lighting design; there are many ways in which sound can pass through light fittings to unwanted areas, and the light fittings themselves may be a cause of sound pollution. There are examples where the fittings have been designed to modify and absorb sound as a functional part of the integration of services within a building.

Another servicing area where lighting design must be considered at an early stage of the design process is that of fire control, since in order to satisfy the structural needs of a fireproof membrane for a ceiling, the light fitting may require to have special fire protection; and in the case of natural light, the design of the building envelope will have to take the requirements of appropriate building regulations into account.

Other services such as partitioning and audio systems may have lighting implications, and in the case of emergency means of escape the co-ordination of lighting with other structural elements is evident.

CONCLUSION

The purpose of this analysis has been to introduce all those factors that need to be considered by the architect at the design stage of any project, factors which are dealt with in later chapters in greater detail.

In projects where there is complexity it will be necessary for the architect to visit all the seven factors identified and to consider the various options in some detail, but in simple structures an acknowledgement that such factors exist and must be considered may be all that the architect needs to know, since in most cases he should be well aware of the sort of solutions that should be applied.

3 Seeing/perception

Seeing has been described as 'to perceive with the eyes' or 'the sense of sight or vision', whilst perception is more widely defined as 'to take in with the mind and senses ... to apprehend ... to become aware of by sight, hearing or other senses'. Whilst seeing and perception may at first appear to be the same thing, the extent to which a space or object is perceived depends first on our sense of sight or vision, but more importantly on the build-up of information from all the stimuli we receive from our surroundings, together with our past experience. For example, we may see the space we are in through the light which falls upon its surfaces, but the way in which we perceive it calls upon a much more developed experience where sight is an important first step, but to which must be added hearing, smell and touch at the sensory level, as well as our emotional and intellectual reactions as formed by our age and experience.

VISION

It is unnecessary to dwell in any detail on the physiological implications of the eye and how it works as many reference books are available on the subject. There are, however, one or two issues which should be borne in mind, as these will affect the way in which a lighting design should be developed.

It is a matter of common observation that up to a point, the more light that is available, the better we see. During the day we tend to place ourselves close to a window to obtain the advantages of daylight, whilst at night we switch on the electric light, and if a means of control is available we set this to provide a light level appropriate to our visual needs. There is a tendency for lighting codes and regulations to concentrate on the level of illumination (illuminance) required to satisfy the needs of vision. The level is clearly important, but this alone ignores our needs for the overall perception of a space.

Visual acuity

It is clear that the more difficult the task to be done, the more light is necessary. Although it is possible to perform at very low levels of light because of the human eye's ability to adapt, if complex work is to be accomplished efficiently such levels would be inadequate – so higher light levels equate with more difficult tasks.

Visual acuity, or the ability to discern detail, is directly related to the illumination level on the object, but relates also to the contrast between the object and its immediate background, or surround. In a work situation, a satisfactory relationship between the task and its background is found to exist when the task light is three times as bright as its background. It should not be forgotten also that brightness relates to the reflectance and colour of the various objects. For example if a task is of low reflectance, the amount of light on it will need to be increased if a 3:1 relationship is to be maintained with its background.

The function of a building will determine the amount of light required. Some buildings or areas of buildings will demand a high level of light to serve the visual needs of the occupants, whilst other areas may be satisfied by lower levels. Suggested levels are found in publications such as the codes of practice of the different Illuminating Engineering Societies.

Examples might be as follows:*

High Bay storage	20 lux
Boiler house	100 lux
The nave of a church	100–200 lux
Living room in a home	100–300 lux
School teaching place	300 lux
Offices, computer stations	300 lux
Offices, general	500 lux
Precision assembly, electrical	1000 lux
Hospital operating theatre (over limited area)	15 000–30 000 Lux

*Taken from the *Code for Interior Lighting*, CIBSE UK, 1994

This abbreviated list shows the wide variety of light levels, from as low as 20 lux to as high as 30 000 lux, required depending upon the difficulty of the task. In addition, people with old sight require higher levels.

When considering the different lighting systems to be evaluated, it should never be forgotten that the chosen system must satisfy both the emotional as well as the visual needs of the occupants. There are programmes where the perception of a space is at least as important as the visual needs of the occupants. An example of this would be a church, where once the visual needs of the congregation and clergy are satisfied so that the ceremony can be performed, how the space appears throughout the day, by daylight and by artificial lighting, becomes important, as it is this that will determine the visual success of the church. Visual performance is only one aspect of perception.

Glare

Glare is the enemy of good lighting. One would have hoped that at the beginning of the twenty-first century no building would be lit by day or by night in such a manner that glare is permitted, yet important buildings still suffer from the bad effects of glare – glare from natural light where the windows have been ill-conceived allowing too great a contrast between the view of the sky outside and the interior surfaces of the building; and glare from the artificial sources within.

It is easier to deal with the problems that result from glare on the drawing board rather than when the building is complete, and for this reason methods of calculation have been devised by the Illuminating

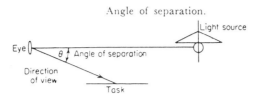

Figure 3.1
Angle of separation. Disability glare is reduced
by reducing the brightness of the light source,
or increasing the angle of separation. (Phillips,
Derek, *Lighting in Architectural Design*, McGraw
Hill, 1964)

Engineering Societies in different countries to overcome the problems. It should be stressed that whilst uncomfortable conditions result from glare, if the factors which cause glare are removed it will not necessarily produce comfortable conditions. Comfort, or the perception which generates it, is much more complicated, and is not only related to the appearance of the space, but also to the emotional and intellectual experience of the individual concerned. The best that can be done where comfort is an important criterion is to ensure that conditions are created which will achieve comfort for the maximum number of people.

There are two basic types of glare: discomfort and disability and whilst the latter is the more important in terms of visual performance, the former may have a greater impression on the perception of a space. It is possible that both forms of glare may be experienced at the same time, and from the same light source, and it is necessary to understand the nature of the glare problem if it is to be remedied.

Both types of glare are the result of unacceptable brightness contrasts. Disability glare results from light sources that are too close to the line of sight, where the brightness of the source is such as to inhibit the view of the task.

A simple explanation is given in Fig 3.1 where if the angle of separation between the seeing task and the light source is too small the eye will be inhibited by the bright light source. This may be the result of daylight coming from a window, but this can generally be overcome by the individual adapting to the direction of flow of light by altering his position, as one would do naturally in the home. The problem becomes more difficult in a static work situation.

To overcome disability glare, it will either be necessary to increase the angle of separation of the light source, or reduce its brightness, whilst maintaining or increasing the brightness of the task.

Another factor that can be altered to improve the situation is the balance of brightness of the surround to the task, and here it is important to ensure that the reflectances of adjacent surfaces are controlled, which is not so much a lighting problem as one of interior decoration.

The example most often cited to illustrate disability glare is that of the headlights of an oncoming car. At night it is necessary to concentrate one's eyes away from the brightness of the headlight, whilst the same brightness seen during the day is quite acceptable.

When the effects of disability glare are fully understood, the phenomenon can be utilized as a design factor to impede the views of unwanted detail, as for example in a retail situation where spotlights are hung below a service ceiling containing pipes or ducts, and where the brightness of the lamps conceals the view above (see Figure 3.2).

Discomfort glare does not in itself interfere with the process of seeing because although it is caused by too great a contrast between a light source, daylight or artificial, with its surround, it may not inhibit a person's ability to discern necessary detail and therefore to perform within the space. Its effect is more concerned with the perception of the

Figure 3.2
An early example of unshielded light sources being used as disability glare sources in a Rotterdam store, to inhibit the views of pipework and ducting above the louvred ceiling. (Copyright International Lighting Review)

space, since an unacceptable contrast between the surfaces or edges caused by an inappropriate relationship with lighting equipment will make the appearance of the space less pleasant; the information of the space gained by the eye in such a case may not conform to the careful interrelationship of space sought by the architect.

Figure 3.3
A window at the Architectural Association building in London, where the reveals are splayed to balance the contrast of light at the window. The splays are used also to conceal shutters. (Copyright Phillips, Derek, *Lighting in Architectural Design*, McGraw Hill, 1964, p.40)

As with disability glare, formulae are available to investigate the discomfort that may arise from the interrelationship of lighting equipment with building surfaces. As already mentioned, discomfort glare can be experienced at the same time and from the same sources as disability. If disability glare is removed by any of the means already suggested, discomfort may still remain and means must be found to overcome it.

Another type of glare to affect vision is that of reflected glare where the image of a light fitting or the light from a window is seen reflected in shiny surfaces. An example of this might be the reflected image of downlighting on a glossy painted wall surface; such glare might not inhibit a person's performance in relation to the wall or corridor on which the reflections occur, but it would impair the perception of the wall.

A more serious example of reflected glare is on the surface of a video screen, where the contrast between the image on the screen and its surface is reduced, diminishing the information a person can derive and resulting in a serious loss of vision. Another example is in any glossy surface where the relationship of the light source and the surface causes reflections and unacceptable contrasts, as is often the case with paintings. It is important that the relationship of the light source and the surface be understood if impaired vision is not to result.

Glitter, or sparkle, is also a form of glare, but one that when associated with light sources, for example in a decorative pendant chandelier, can be acceptable. Ideally one would not place such a chandelier in a work situation.

EMOTION AND INTELLECT

Those factors so far discussed are concerned with vision, or the functional aspects of lighting design, originally described as 'enough light, where you want it, how you want it.'[1] These aspects of lighting are critical to the success of a building, but the necessary qualities to satisfying the human needs of emotion and intellect are concerned with the perception of a building experienced as a whole rather than our ability to see; they are less easy to quantify.

Our concern here is with the intangible aspects of lighting, which is why there is little likelihood of the computer taking over lighting design. It requires first an understanding by the architect of what the building is to be about and then a symbiosis with the lighting designer to ensure that he or she understands what it is the architect is trying to achieve. It is only when the architect fully understands the emotional response he requires and can convey this to the lighting designer that the best results will be achieved.

This will come from experience of other buildings and a knowledge of history – not always something which can be acquired in the training that an architect receives. It comes from the observation of the buildings of past history, buildings which have delighted generations of people, buildings which have a 'classic' quality derived from a unity of experience, present as often in the small chapel as the large cathedral (see Figures 3.4 and 3.5).

These examples have been taken from a single architectural programme, that of a church, but the principle could equally well apply

---

[1] Phillips, Derek, *Lighting in Architectural Design*, McGraw Hill, 1964, p. 45.

Figure 3.4
The daylighting at the cathedral of Sta. Sophia in Istanbul shows a sophistication of design, likely to have been assisted by model studies. (Photographer Derek Phillips)

to others, such as a house, hospital, library or art gallery. It is the experience of the unity of the building that will have an appeal, not the copying of architectural detail. What is required is an understanding of the subjective response to the earlier buildings which will lead to an understanding of the new.

Without in any way denying the influence that artificial sources of light can bring to the emotional qualities of a building after dark, it is important to mention those aspects of the natural environment that contribute to its unity during the day.

1   The direction of the light, which provides modelling to the interior, the nature of sun and sky. It is this direction which provides the clarity of structure with which unity is expressed.
2   The contact with the exterior beyond, such as a view through the window, an experience of the weather and the world outside.
3   The natural colour associated with daylight which imparts reality to the interior.
4   The mood created by the variation of light, from day to day, and time to time as affected by the weather and seasons. The dynamics of lighting.

Figure 3.5
Zwiefalten Abbey in Bavaria, where the structure of the nave is designed as giant vertical louvres to modify the daylight entering the building. (Copyright James Bell)

The human aspirations of the building at night will differ from those during the day and it would be wrong to try to turn night into day even though many theories of lighting design have been based on this concept. Even the great Waldram in his designs for the lighting of Gloucester Cathedral in the 1950s felt the necessity to design the artificial lighting to echo the flow of light in the nave from South to North which daylight produced during the day.

Artificial light has its own quality, and although during the day there may be good reasons to supplement the natural source, for example, to increase the amount of light far from a window, after dark it will need to do more than satisfy the needs of vision; it must also contribute to those other aspects of clarity, colour and stimulation no longer provided by daylight, but which add to the unity of the space during the day.

It is worth looking at the four characteristics of daylight mentioned below – not to copy them – but to see to what extent the emotional needs they fulfil may have a resonance in night time lighting.

Clarity of structure

The provision of the required illumination level will have little to do with the way in which the spaces are modelled by light after dark, and the appearance of a space during daylight is not the only way in which it may be perceived; the architect may wish to add a different unity more appropriate to the way in which the emotional impact of the building is seen as relevant at night. For example, during the day the architect may wish the building to provide a light and airy environment, whilst at night it may be felt that a more dramatic impression would serve the needs of the building better. We react differently to the atmosphere of a space after dark, and theatre designers, for example, have been able to create alternative conditions within the same space. Perhaps this is why theatre designers have in the past been some of the most successful interior lighting designers.[2]

Contact with the world outside

After dark it will be more difficult to experience the day and the scene outside, but where appropriate suitable contact can be made with views out to a well-considered scheme of exterior lighting, as for example in the

Figure 3.6
A simple scheme of outside lighting provides an extended view to the garden beyond at night. (Photographer Derek Phillips)

Figure 3.7
The view from a city boardroom over the river Thames in London with the lighting provided by floodlit buildings in the distance. (Photographer Derek Phillips)

[2] Abe Feder and Howard Brandston in the United States, and Andre Tammas in the United Kingdom.

view through a window to a flood-lit garden in a domestic situation. This provides an extension of the living space, creating its own emotional response (see Figures 3.6 and 3.7).

Natural colour

The colour of light is always of importance, whether during the day or at night. It has even been suggested that good colour can allow a lower illumination level. This seems to be sustained when applied to a building after dark. Here the colour may not be the colour of daylight because we tend to accept a warmer colour at night.

In those areas of a building where no daylight penetrates there may be a case for a change of emphasis from day to night to answer the expectations of a different atmosphere after dark. We react differently to our surroundings at night, and an example of this might be where the artificial lighting for an underground shopping arcade reflects the change after dark.

Figure 3.8

Figure 3.9
An underground shopping arcade at Sha Tin in Hong Kong where the lighting is changed from day to night. During the day (see Figure 3.8) fluorescent lighting is used whilst at night filament sources are employed (see Figure 3.9), both above a louvred ceiling. (Photographer Derek Phillips)

In this example, the cooler fluorescent lighting which might be expected during the day changes to a filament installation at night. All the light fittings are placed above a louvred ceiling so that only one method of lighting is visible at any one time (see Figure 3.9).

Variety

Clearly there are some aspects of daylight, such as its variety, that cannot easily be translated into the night-time response. Where such changes have been recreated on some mechanical basis it is more in the realm of theatre and not always appropriate.

This however does not preclude changes of atmosphere appropriate to the needs of a space. Variety of artificial lighting is wholly appropriate to the change in emphasis of the different functions of a single space in a building (the difference in the working area to the walk-through area of an office) or the different spaces themselves: the corridors, reception areas, leisure spaces and so on. It is this variety, leading to a sequential experience within a building, which is important, so variety is not inimical to the artificial scene, it is just that it needs to be planned with care and not result from chance.

It was stated at the outset that the intangible aspects of lighting design which give rise to our emotional responses cannot be quantified but result from an architect's knowledge and experience. It is therefore important that the lighting designer should learn to understand and appreciate the aims and expectations of the architect, or as Waldram described them thirty years ago, those aspects which are 'beyond engineering.'

AGE AND HEALTH

The published illumination levels are designed for people of average sight, and it is generally agreed that as sight deteriorates with age a higher illumination level is required by the 'old' eye to achieve the same level of visual acuity as the 'average.' It would be uneconomic to increase the overall level of light in a building to satisfy old sight. In any event the use of glasses can increase the visual acuity of those concerned to acceptable levels. It is only where a building has been nominated for the special use of old people that account should be taken. In this case special care is needed to provide higher levels of illumination at points where this may be required, at reading or cooking locations for example. But what is more important is that the architect plans for exceptional clarity of form at staircases and other danger zones.

One of the greatest dangers to old people is that of misreading heights and distances so that they may fall, and much can be done by the architect and lighting designer to ensure that all the necessary visual cues are available to make the building safe. It is this rather than more overall light which is important.

A deterioration of vision due to ageing is a natural phenomenon unaffected by the quality of the lighting, but there are aspects of lighting which do have an effect upon health. Most of these have already been covered earlier, in terms of glare and the stress that this can cause, which, if not rectified can lead to strain and adverse health conditions.

There is little doubt that where the working conditions are poor, the levels of light inadequate, and where contrasts are such as to have an adverse effect upon vision, it can cause serious health problems. In the early days of fluorescent light, many complaints were made by staff who suffered from headaches and other disorders. This was largely due to the presence of flicker at the ends of tubes, particularly from the early halophosphor lamps, which could be seen by the eye. This is less prevalent today with the new triphosphor lamps and has been entirely avoided by the new high frequency circuits available. Flicker, however, remains a problem with some of the high-intensity discharge sources such as metal halide, where its effects have not been entirely eliminated. Whilst flicker may not prove to be a serious problem in many architectural programmes, as for example in a retail situation, in a fixed work location it should be eliminated.

A problem often associated with lighting, described as Seasonal Affective Disorder (SAD), affects quite a large number of sufferers (a figure of a million has been quoted) and is said to derive from the lack of sunlight during the winter months, from September to April. Whilst for most people this is not a serious problem, since most of us feel slight depression when there is a lack of sunlight over a long period, for others it can be debilitating.

It is certain that those people in work situations which have access to natural light, with periodic sunlight, are likely to be less affected by SAD, whilst those in totally artificial conditions are more likely to suffer. What is clear is that whilst a building should provide access to natural light where possible, it would be quite impractical to try to overcome the effects of SAD by increasing the normal building lighting system.

The health of a workforce is vital to the success of an organization, and it is therefore of the greatest importance that the lighting of a building acts with other environmental systems to create healthy conditions. To cut down on lighting to save money at the outset of a building design makes no sense if it reduces the performance of the much more expensive work force. Lighting design can contribute to the health and amenity and consequently to the performance of the whole operation.

4 Daylight

THE IMPORTANCE OF WINDOWS

No one can deny the historical importance of daylight in determining the form of buildings since, together with the effects of climate and location, daylight availability was fundamental to their design. However, with the introduction of modern sources of electric light, and particularly because of their increasing efficiency since the Second World War, by the 1960s the need to introduce daylight into buildings had appeared to diminish. A number of architectural programmes such as offices, shopping centres, factories, sports buildings and even schools were developed as 'blind' or 'semi-blind' boxes on the assumption that other environmental factors such as heating, cooling and acoustics would be better served if there were no windows – the best window was no window.

This assumption was engineering biased, where all the less tangible advantages of daylight entry to a building could be ignored. It was never a sound argument and is even less so today. The introduction of daylight with all its variety has always been recognized by architects as having positive advantages, and now this view has gained ground due to the realization that our finite resources of energy must be conserved in world terms. The developed nations need to consider how savings of energy through building design can make a positive contribution.

A brief history of the development of daylight design

It is helpful to start with a brief review of daylighting in domestic buildings to illustrate how this type of architectural programme was informed by the need to admit daylight when artificial lighting as we know it today was not available, and to show how this influenced the exterior appearance of buildings.

Leaving aside the rock and cave dwellings of primitive man, it is the Roman courtyard house which typifies the early development of domestic architecture, illustrating the role played by daylight in establishing its form and taking into account the influences of climate and culture.

The Roman house needed to provide shade to combat heat whilst at the same time delivering light to the interior. The form developed was a courtyard plan with an entrance atrium. This form, upon which our modern atria are based, was enclosed on all sides by buildings with roofs sloping to a columned peristyle or walkway around a courtyard. Rooms would generally be lit from large doorways off this courtyard, or from

small windows to the street or garden beyond. The problem here was not so much to admit maximum daylight as to reduce solar gain to an acceptable level – the effect of climate upon form (see Figure 4.1).

The design of the Mediaeval house in England responded to the feudal social structure as well as to a different and colder climate. Houses were often dominated by a large communal hall. For climatic reasons daylight openings were generally of restricted size and controlled by the use of wooden shutters or by a variety of translucent materials such as mica, parchment or oiled linen. The use of glass became more generally available by the fifteenth century, allowing large expanses of weather-sealed glazing. Windows were essentially asymmetrical, being located to satisfy the needs of the interior, and there are many examples of such houses where the interiors are beautifully modelled by daylight, gaining immeasurably from the changing variety of light available (see Figure 4.2).·

Figure 4.2
A mediaeval hall lit during the day by unglazed windows, shutters being used to control the climate. At night the hall was lit by candles and light from the central fireplace. (Weald and Downland Museum, Singleton. Photographer Derek Phillips)

The Renaissance in Italy saw a formalization of the window pattern within symmetrical façades, often paying little regard to the interior spaces behind. Whilst the long façades of the palaces of the merchant princes gave an impression of grandeur and solidity, the needs of daylighting dictated that the plans were limited to 15–20 metres using courtyards or gardens at the rear.

Such bilateral daylighting was not dissimilar to that available in the traditional office block today. The windows were generally vertical, and with tall ceilings this ensured maximum penetration of light, which with building height limited to two to three stories, prevented the overshadowing of other façades.

The Queen's House at Greenwich, designed by the architect Inigo Jones in the seventeenth century, saw the development of the toplit internal room which allowed deeper plan buildings. A well documented example of this is Keddleston Hall (see Figure 4.3), whilst perhaps the best-known house, and one that had a significant influence on the houses in England, is the Palladian Villa at Vicenza, from which Chiswick House was derived, which is based around a central dome admitting daylight (see Figure 4.4).

Figure 4.3
Daylight entering the saloon through the dome at Keddleston Hall, allowing a deeper plan. (Copyright National Trust)

The eighteenth century saw much refinement of window detailing. One of the important aspects of window design of the period was the manner in which the light was graded into the interior, to reduce the harsh contrast between the brightness outside and the brightness within – a lesson well learnt many years before, as illustrated by the curved window

Figure 4.4
Chiswick House exterior showing the central dome admitting daylight. (Photographer Derek Phillips)

embrasures at the Château de Chenonçeau in the Loire, which introduces daylight into the bridge, articulating the pattern of light and shade and leading the eye forward (see Figure 4.5).

The house of the architect Sir John Soane (now the Soane Museum) in Lincolns Inn Fields in London built in the late eighteenth century epitomizes the use of daylight in domestic architecture. Soane introduced light into the interior from exterior walls by a subtle grading of the light, by using deep chamfered reveals to vertical windows; whilst in order to admit daylight into a deep interior he adopted the use of a wide variety of domes and other forms of overhead daylight.

Figure 4.5
The bridge at the Château de Chenonçeau in the Loire where daylight is introduced to the interior by the curved window embrasure, reducing the contrast between the inside and the outside. (Photographer Derek Phillips)

The narrow-fronted houses along the canals of Amsterdam with their high rooms and large windows, represent a further example of the importance given to the quality of daylight. Painters such as Vermeer were greatly influenced in their work by natural light, modifying and directing it by simple controls such as shutters. The Dutch have a great understanding and love of daylight, as can be seen not only in their domestic architecture but also in the design of their great churches (see Figure 4.6).

Figure 4.6
The large windows of merchants' houses along the canals of Amsterdam indicate the importance of daylight. (Photographer Derek Phillips)

More recently the structural revolution of the modern movement in the 1930s allowed wrap-around and strip windows which filled the whole frontages of houses, as in the Connel Ward and Lucas house at Moore Park, a celebration of sunlight and view. The Maison Veere in Paris, with its whole walls of glass brick, illustrates as no other the fascination for glass and daylight of the period (see Figure 4.7).

This short historical introduction has been based for simplicity on domestic architecture, but it should be sufficient to emphasize the importance of daylight in all interiors and show how the structure of buildings has been influenced by the external environment in different climates and at different periods of history.

There is a close correlation between the interior of the building and its exterior appearance as illustrated by the Ismaili Centre in Kensington, London (see Case Study 58 Ismaili Centre). This shows the connection between the interior function of the spaces behind the façade and the exter-

Figure 4.7
The house at Moore Park designed by architects
Connell Ward and Lucas in the 1930s illustrates
the way in which new developments in structure
allowed wrap around windows to increase the
amount of daylight. (Photographer Derek
Phillips)

nal expression or appearance. At the first floor level the glazing gives both
light and view to a social area, whilst minimum slit windows relate to the
prayer hall above, providing a low level contemplative atmosphere.

THE UNIQUE QUALITIES OF DAYLIGHT

The unique qualities of daylight and sunlight make their introduction into
buildings as relevant as when there was no viable alternative in artificial
sources. These intangible factors which have shaped man's development
cannot be overestimated, and it is always useful to remind ourselves of
them:

Change and variety
Modelling and orientation
Sunlight effect
Colour
View

Change and variety

Daylight is a constantly changing source, varying from time of day,
season of the year and the state of weather, whether sunny or cloudy. Far
from being a disadvantage it is this variety which provides a dynamic and
appealing appearance to an interior.

Changes in the weather can be observed, modifying the appearance of
a room and helping us to react to the external environment during the
day. This is a different measure of experience from being in a theatre
when you only discover what the weather is like on going outside. The
experience of the first fall of snow, with light reflected upwards to
ceilings, provides the interior with another facet of its changing appearance, as exciting as the first sunlight after a period of dull days.

The human body has a capacity for adaptation, particularly in the field
of vision, with a need to exercise this response. Perception itself depends

upon a degree of change; the appearance of our surroundings alters with time, and if we have confidence in their continuing reality, it is because changes in their lit appearance allow us to continue an exploration of their character. Change and variety are at the heart of daylighting, as is the medium through which it is delivered, the window.

Appreciation of the temporal quality of daylighting cannot be better expressed than by one of America's leading architects, Louis Kahn, the architect for the Kimbell Art Museum in Texas: 'I can't define a space really as a space, unless I have natural light . . . natural light gives mood to space by nuances of light in the time of day and the season of the year, as it enters and modifies the space' (see Figure 4.8).

Figure 4.8
The use of concealed daylight by the American Architect Louis Kahn in the Kimbell Art Museum, Texas, is typical of his work. Whilst the paintings are generally lit by artificial light to the accepted standards for conservation, the ambience of the spaces reacts to the daylight outside so that visitors feel they are entering a daylit space. (Copyright Kit Cuttle)

Figure 4.9
The deep, single-storey building at Stansted Airport allows the whole interior to be flooded with daylight from carefully designed roof lights. (Photographer Derek Phillips)

Modelling and orientation

Daylight has direction because of the movement of the sun from morning to evening, varying as it does from season to season and with higher brightness at different times of day. This direction adds both orientation and modelling to the appearance of a space; it provides clarity.

We perceive and understand the spaces we inhabit by their three dimensional quality. Daylight provides changing modelling and increases our perception of time and space, with the added aspect of orientation, giving a person an awareness of where he is in relation to the exterior (see Figure 4.9).

Modelling

Daylight modelling, or the emphasis provided by shadow patterns on surfaces and form resulting from the direction and flow of natural light, is related to building orientation, coupled with the nature of the windows or means of daylight entry. The appearance of the interior architecture is determined by the physical surfaces, edges and textures when acted upon by the light falling on them.

Interior spaces are judged to be pleasant, bright or gloomy as a result of modelling effects. Horizontal modelling from side windows is the general experience, as affected by the window embrasure. The modelling on a dull overcast day is a different experience from that of a sunny day. Light from windows on adjacent walls assists in modelling, with the main direction of light being supplemented by reflected light.

Modelling from overhead daylight is of a different order, with greater vertical emphasis, and is useful in art galleries displaying sculpture, where the nature of the modelling more closely resembles that from the daylight outside (see Figure 4.10).

Orientation

The importance of orientation is acknowledged in the setting of the building on its site and its relationship to the sun path to achieve the optimum natural lighting solution for the building's function, whilst a knowledge of the world outside assists an individual's understanding of his whereabouts within a building.

It will not always be possible to provide the optimum orientation for a building on its site, or its best relationship with the sun path, for example where a building is set into a rigid street pattern or where there is overshadowing from a neighbouring building, but the question of orientation should always be a consideration. The best solution should be simple enough to achieve on a greenfield site, when there is no excuse to get it wrong. When in doubt it is useful to model different solutions to test them out.

A good example of this is shopping centres. Victorian examples permitted daylight to enter through roof lights, providing necessary environmental information but the role of daylight was misunderstood in some of the early shopping centres in the 1960s which were built as blind boxes. The latter were not liked, and it is unlikely that such solutions will be repeated today. People enjoy shopping during the day in spaces lit well by daylight, whilst at night it is accepted that there can be a complete change of atmosphere, created by artificial conditions (see Figure 4.11).

Sunlight effect

Where possible the orientation of the building will have optimized the entry of sunlight; this increases the overall level of light, and assists in providing other environmental factors mentioned, such as change and

Figure 4.10
Michaelangelo's David statue lit from a dome roof light allows the modelling to change from day to day and hour to hour. (Photographer Derek Phillips)

Figure 4.11
The overhead daylighting at the Bentalls Centre in Kingston provides very adequate natural light during the day to all the public areas, whilst the shops themselves, entered off this central space, are artificially lit. (Photographer Derek Phillips)

variety, whilst ensuring delight – the delight, for example, experienced when getting up to a sunlit day. This is not something that can be measured, and is therefore difficult to assess by engineering formulae. When sunlight is available, there is a human need for it to be taken into account, and a sense of disappointment when this is denied.

Sunlight is to be welcomed in buildings and often provides a positive architectural element, for example when shafts of sunlight are directed through windows at high level, as in gothic cathedrals. It is equally important in small scale programmes such as in the home. Sunlight is fundamentally good both therapeutically and visually, but it is always important to consider the obverse side in terms of heat gain and glare. Control may need to be exercised over the sunlight entering a building, depending upon geographical location and the purpose of the building.

There are some programmes, such as art galleries, where for conservation reasons sunlight needs to be excluded, but even here the effect of sunlight may be obtained without its energy affecting the works of art. A view from a north-facing window towards a sunlit scene provides a substitute for the lack of sun inside (see Case Study 49 The Burrell Collection). The question of sunlight control is dealt with later, but there are clearly work situations where the effects of direct sunlight would be disadvantageous (see Figures 4.12 and 4.13).

Figure 4.12
A sunlit staircase at council offices in Winchester makes the impression of sunlight and the view beyond visible from the offices despite the fact that no sunlight reaches the offices themselves. (Photographer Derek Phillips)

Figure 4.13
Daylight and sunlight in a shopping centre in Watford give the appearance of an external scene, whilst giving protection from the weather. (Photographer Derek Phillips)

Colour

Man developed in natural conditions and so we instinctively believe the colours of objects seen under daylight. Whilst the colours produced by artificial light sources have greatly improved and do not produce the distortions associated with early electric lamps, we still take things to the light to confirm their true colour, and by 'the light,' we mean daylight.

Daylight is the colour reference, since all other forms of light change the perceived colour to a greater or lesser degree. Daylight is thought to be the true colour despite the fact that it varies in hue from morning to evening and is enhanced by sunlight. This is of particular importance in buildings where we have learned to adapt to the natural change which occurs and a white surface still appears white in evening light.

In the nineteenth century large department stores encouraged daylight entry for a variety of reasons, not least because of the advantage of 'real colour appearance.' However, the practice was discouraged in the twentieth century, partly to allow multiple floors of shopping area and additional wall space for displays. This led to blind boxes, which not only produced artificial conditions within but unhappy results externally as dull cityscape.

With the present renewed interest in daylighting, some large stores have been designed to introduce daylight, in a low energy ethos. Here the feeling of well-being of shoppers is enhanced by the knowledge that the colour of goods and materials on sale are their natural colour. There is an additional benefit to those who work and spend long hours inside the shop.

View

Windows – through which daylight is introduced to the interior, where the light is modified and controlled, and from which the views out beyond the building are obtained – are at the heart of the matter. Windows throughout history have often determined not just the appearance of the exterior but the whole form of the building. Similarly, the absence of windows dictates the exterior appearance, emphasizing the artificiality of the interior.

The view out through a window or how we perceive the world outside is a dynamic experience associated with changes in daylight, sunlight and season. At its lowest level a view satisfies man's physiological need for the eye to adapt and readapt to distance by providing an extension of view and stimulating an awareness of the environment beyond the building (see Case Study 40 p.183).

The quality of the view is clearly of importance. Some views are of exceptional beauty and provide inspiration and delight, but any view that extends the experience of the world beyond, however banal, is better than no view. Research in Pennsylvania suggests that patients in hospital recover more quickly where there is a view. (Uhlrich, R. (1984). View through a window may influence recovery from surgery. *Science*, **224**, 420–1.) Another example of the benefits of a view are school classrooms. Originally it was thought that preventing children from having a view out of the window would ensure greater concentration on their work, now a view is regarded on balance to be stimulating to the learning process as well as having many other advantages.

In the 1960s many windowless buildings were designed, either to overcome adverse climatic conditions, as in windowless schools in West Virginia, or for reasons of internal space planning. There are clearly some

Figure 4.14
The view out to the garden of this house in Chipperfield was clearly uppermost in the mind of the architect Maxwell Fry in his designs for both daylight and view. (Photographer Derek Phillips)

Figure 4.15
The interior of this school in West Virginia suffers from lack of both daylight and view, built at a time when neither was considered to be of importance. No attention to well-designed artificial lighting can make up for their lack of amenity. (Photographer Derek Phillips)

Figure 4.16
In Hampshire the schools relate well to the natural environment and enjoy the advantage of well considered daylight. (Photographer Derek Phillips)

architectural programmes where the main element of accommodation, as in a theatre or cinema, demands an absence of daylight. However for the vast majority of architectural programmes, a view is important. Lack of any contact with the outside is a form of deprivation (see Figure 4.15).

The quality of view will depend not only on the exterior but upon the location, size, shape and detailing of the window itself since however stimulating the exterior might be, the view out will be inhibited where the windows are too small, break up a view, or are at the wrong height.

Throughout history the window has been associated with a sense of place, for example in the seats built within the thickness of walls in early fortified houses to the bay window associated with a window seat in a modern house or apartment. Window seats are regarded as an attractive feature of a house, and there are many other architectural programmes, such as leisure buildings, where seats associated with windows have been adopted (see Figure 4.16).

Where it is inappropriate to associate a window with a sitting position, windows should still be associated with the view as the general supposition must be that a view is good. The adaptation of the eye to distance, an extension of view and an awareness of the environment beyond the building, all contribute to man's well-being. These unique qualities of daylight have been exploited throughout history and it is when these qualities are disregarded that buildings fail to provide satisfaction (see Figure 4.15).

ENERGY

Thermal comfort

Windows are closely related to the problems of energy use in buildings. In summer the heat gain from the entry of sunlight through south-facing façades can cause a problem of over-heating whilst in winter the extra heat can be useful. It is necessary, therefore, when assessing the visual advantages of sunlight penetration discussed earlier to be aware of the energy considerations for the different façades of the building at different times of year.

The modern movement in architecture of the 1930s rejoiced in the rediscovery of daylight, but the introduction of more light brought its own problems, of overheating and cooling, poor noise control, and lack of ventilation which has tended to lead to air conditioning and its contingent use of energy.

In the 1990s architects have learned to overcome the environmental disadvantages of daylight and are learning to deal with them with hi-tech solutions. It is clear that the future lies in the development of types of high performance glass, variable transmission glass and other means of solar control.

Glass in windows has a crucial role to play in improving the internal environment, which, associated with the need to reduce energy in buildings, has led towards new window designs, where the associated problems of ventilation, solar gain, glare and noise pollution suggest an integrated solution.

Glare

Glare through windows, like glare from artificial light sources, results from contrast, contrast between the light source, both skylight and sunlight, and the window wall when viewed from the inside.

Direct glare will be experienced through west-facing windows in the northern hemisphere in winter from low-angled winter sunlight. But glare is not confined to this obvious example. Views of the sky may provide an unacceptable contrast with the interior surfaces of the room, and much can be done by window design and window reveals to ameliorate the high contrast and associated glare (see Figure 4.5).

There is a correlation between solutions to the control of sunlight for thermal reasons and for those of glare – cutting out sunlight from different directions to avoid overheating in summer will reduce glare for the interior – but it should be borne in mind that the reduction of glare may not by itself provide a solution to the provision of thermal comfort.

The avoidance of glare is a maximum priority for most architectural programmes, particularly those with fixed work positions, and 'add-ons' after the building is complete, such as internal shading devices, are not the solution. Much can be done by external shading or high-tech glass; what is important is that glare avoidance is an integral part of design strategy being planned for and executed at the design stage of the building.

Noise

The transmission of noise from the outside to the inside of a building is a particular problem associated with sites in the centre of cities or close to airports or motorways, and since noise penetration and window openings are closely related, solutions need to be found in the acoustic design of the window and its associated framing structure.

Where windows are designed to open they must when closed be capable of suitable noise reduction (attenuation). When the windows are permanently closed, as may be the case in air conditioned buildings, the window design must reduce sound entry to an acceptable level. Airborne sound finds its way through any gap, and so some form of flexible sealing system will be required at vulnerable junctions.

Much can be achieved by the choice of glass and various solutions are available, such as double or triple glazing, wide cavities with insulating boundaries and different glass weights. The type of glass is related not only to problems of noise, but also to thermal and visual considerations, so much thought must be given to its choice.

Structure

Windows are a basic element of the fabric of a building, so that in addition to the visual qualities inherent in their design, the problems of structural strength and stability need to be addressed.

Windows are the first line of defence against the local climate, and where this is adverse in terms of temperature, as for example in North America or Scandinavia, windows have been developed over the years to keep out the severe cold. In the more temperate climate of Britain it has been less necessary to develop window designs to overcome the extreme rigours of winter, but each climate will set its own parameters, and as expectations rise, so will the need to solve climatic problems at the window become more exacting.

Windows need to cope with the maximum wind pressure that can be expected in a particular location, and the possible threat of missiles and vandalism may require the use of high-impact resistance glass. The window may also have to provide some protection against fire.

On certain elevations the window may require protection from solar radiation, either through the selection of the glass used or by external shading, or both. The use of internal shading is less efficient for thermal control, but is more easily managed. When used, external shading becomes a structural element and is both visually and structurally important: visually it has an impact on the external appearance of the building and structurally it must withstand all the external pressures applied to it.

With the greatly increased use of glass in architecture, where whole walls of glass may be used, the structural aspects of window design become paramount.

The increased size of windows brings up the need to solve the problem of keeping the glass clean, for whilst smaller windows always presented a problem in the past, it could be more easily solved. Glass needs to be maintained and kept clean to ensure that its light-transmitting qualities are maintained and that energy is not wasted. It is necessary to ensure that the means are available for easy access to the glass surfaces both externally and internally, because if this is not considered within the initial design concept it may be difficult to provide them at a later stage.

Overhead daylighting from domes or laylights to internal spaces within a building, as in atriums, presents its own maintenance problems and demands special solutions.

Conclusion

Art and science together could produce a sustainable architecture for the twenty-first century. The original requirement for windows was to keep out the rain and the cold which was thought to be satisfied by a single skin of glass. But now that the artificial lighting energy load is significant, anything that can be done to reduce this by natural means will help. Heat recovery through window design, for example, reduces the heat load on the building. This type of solution related to the thermal design policy for the building will be the key to solutions in the next century, since it may eliminate the need for air conditioning, the window wall becoming the total energy control for the building.

WINDOW TYPES

Vertical windows

Windows set into solid walls where the height is greater than the width are called vertical. These were the most usual form of window. They give horizontal emphasis to the modelling of the interior and the amount and penetration of the available daylight is dependant upon the depth of the space and the ceiling height. The light distribution pattern is affected by the number of windows in the room and their spacing.

The light from vertical windows can satisfy all the unique qualities of natural light in the modelling they provide to the interior surfaces of the space, whilst other qualities, such as variety and change, are already inherent in the introduction of the natural source. Vertical windows allow views out, provided the cill level is at a suitable height, although the horizontal extent of the view will be broken up by the spaces between the windows when seen from the rear of a space.

The detailing of vertical windows achieved its peak in the eighteenth century. Deep-splayed reveals and narrow, tapered glazing bars reduced the contrast between the inside and the outside, so reducing the effect of glare. Window cills were generally low and heads high to maximize the distribution of light into the space. The splayed reveals often acted as shutters for security (see Figure 3.3).

Horizontal windows

Horizontal windows are where the window forms the greater part of the elevation at each level, divided only by the floor spandrel. They are common today in multi-storey buildings, where the frame structure gives flexibility at the window wall. In earlier buildings the load bearing structure would have made this impossible. Horizontal windows were developed to provide daylight for early workshops in Britain in the late eighteenth and early nineteenth century. Typical examples are the Lancashire loom shops, where crosswall construction coupled with load-bearing mullions allowed continuous window ranges (see Figure 4.17).

Figure 4.17
These early nineteenth century loom shops in Lancashire provide an early example of horizontal windows developed to solve the problems of the industrial revolution. (Copyright W. John Smith)

Figure 4.18
A daylit gallery area at Alvar Aalto's Aalborg Museum, a building that uses overhead daylighting for some of its galleries. (Copyright David Loe)

By the late nineteenth and early twentieth century a new freedom was provided by the frame construction of the Chicago skyscrapers, where the exterior walls of buildings were no longer load-bearing, freeing up the window design and leading on to the modern movement in architecture of the 1930s where horizontal windows wrapped around the corners of buildings.

The natural light from horizontal windows in these buildings provides adequate daylighting for a depth related to the ceiling height in unilaterally lit spaces, and where artificial light is added towards the rear of deeper rooms, a satisfactory solution can be found for work areas. Another alternative is bilateral daylight but the planning constraints that may be implied should be borne in mind. In taller spaces, such as churches, schools or hospitals, a row of horizontal windows at high level (clerestory) which give deeper penetration of daylight and light to the ceiling might be associated with windows at low level satisfying those other environmental needs such as 'view'.

The Window wall

A natural extension of the horizontal window is where the window takes over the whole perimeter of the building and the wall becomes the window. Whilst this form of window can, with suitable controls, both provide adequate natural light and satisfy the environmental needs of the occupants of the space, there are inherent dangers.

Floor to ceiling glazing may have a great attraction to architects in terms of the simplicity of the exterior appearance of a façade, but care needs to be taken to ensure that the occupants' feelings of security are satisfied. In multi-storey buildings, knowledge that the glass between you and the outside is strong enough to withstand any impact may still not overcome an emotional feeling that you might fall off the edge.

There are good structural reasons for there to be a visual break in the glass at the spandrel at each level. In addition, the cill level is an ideal place for the provision of other environmental controls. It is now possible to support very large panes of glass and this has made the window wall

Figure 4.19
An example of bi-lateral daylighting at the Westgate school in Winchester, where the building cross-section has been developed in conjunction with the daylighting design. (Photographer Derek Phillips)

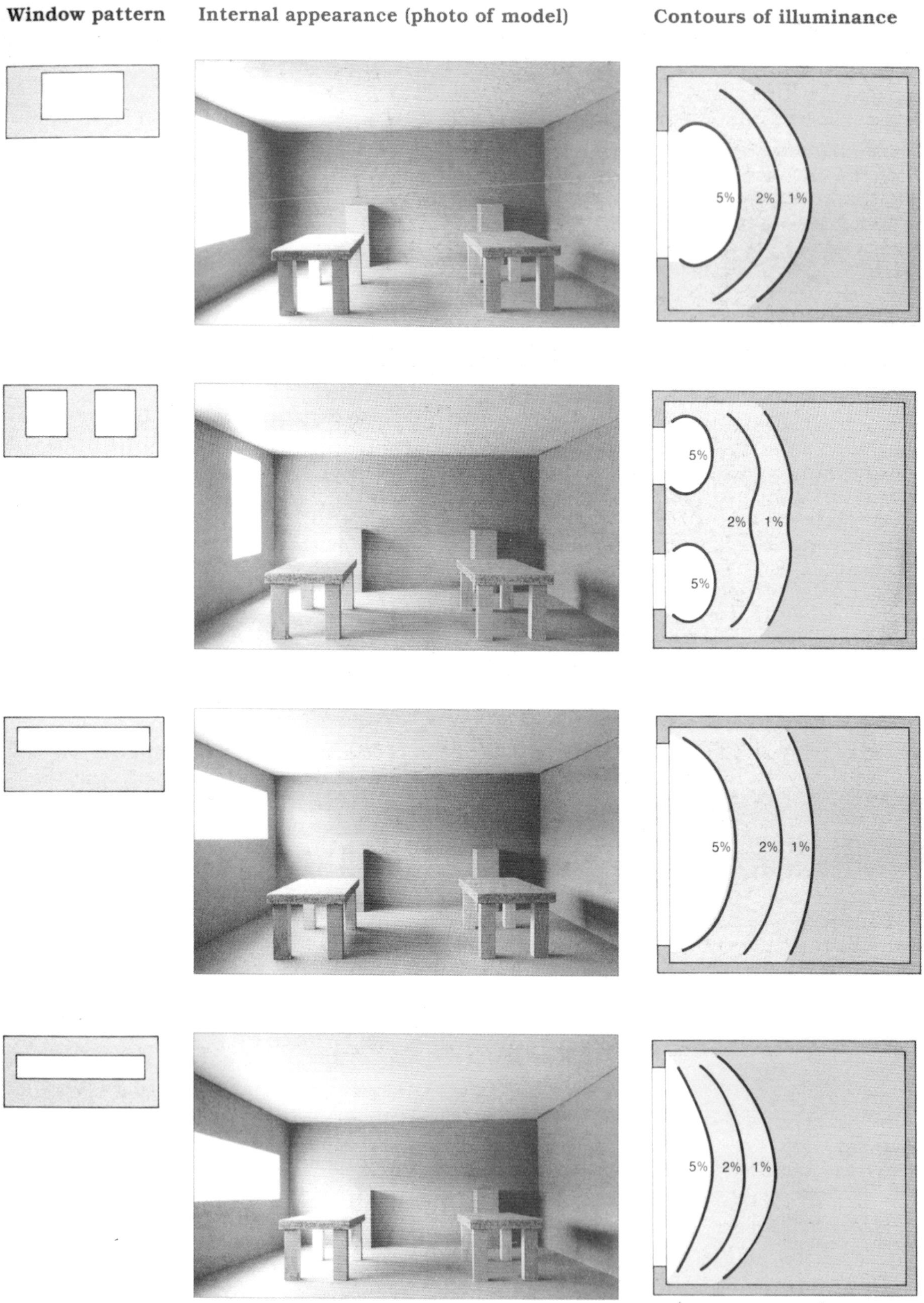
Window pattern
Internal appearance (photo of model)
Contours of illuminance
5%
2%
1%
5%
2%
1%
5%
5%
2%
1%
5%
2%
1%

Figure 4.20
The effect of window patterns on light distribution and shadowing, taken from a series of windows where the window area in each case is the same. (James Bell and William Burt, *Designing Buildings for Daylight*, CRC Ltd, 1995, p. 56)

Figure 4.21
The entrance to Liverpool Street Station where a structural glass wall has been used to stress the verticality of the opening. (Photographer Derek Phillips)

feasible. It has become an important part of the architect's vocabulary, but he or she must ensure that the emotional needs of the occupants are met (see Figure 4.21).

Overhead windows

For deep-plan buildings, or those where perimeter windows may be thought to be inappropriate as, for example in an art gallery, the use of overhead glazing provides a solution. There are many types of overhead glazing, from the domes used to admit light into the interiors of deep-plan historic houses to the great shed roofs of railway stations and the atrium designs providing light to the interiors of our modern shopping centres or office complexes. Where factories require very large spaces to accommodate a manufacturing process, daylight can be provided in single storey buildings by forms of overhead lighting, such as monitor or saw tooth roof lights. The beneficial effects of a daylit environment are obtained despite the fact that some environmental advantages such as view are lacking.

Overhead roof lights bring their own problems, such as access for cleaning and 'heat gain/loss', which have to be addressed, but they give an excellent opportunity to provide a flexibility of plan form in building, whilst providing daylight with more of a vertical emphasis (see Figure 4.22).

Concealed windows

Concealed windows are perhaps not a type on their own as any form of window, horizontal, vertical or even overhead, can be concealed from critical viewpoints, but they constitute a method of daylighting using light from sources which cannot easily be discerned or where the nature of the window opening admitting light is screened by structure.

Concealed windows are a particular feature of church architecture, designed to provide emphasis towards the altar end. Baffled by the structure of the church, they are concealed from the normal views of the congregation (see Case Studies 8 Bagsvaerd Church and 5 Clifton Cathedral). The baroque churches of Southern Germany employed such techniques to advantage, and the concept has been applied to present day architecture with great effect where artificial light sources from the same locations are concealed from view, eliminating glare (see Figure 4.23).

Conclusion

The way daylight is introduced into a space through the window detailing must be considered carefully in terms of its visual impact, as well as other environmental and structural factors. Careful window detailing is the key to acceptable contrasts.

Where ceiling heights are low, as in most modern office buildings, the daylight penetration will be such as to require supplementary artificial light at the rear of the space, unless further daylight can be arranged on the side opposite to the window, or the room depth kept to a minimum. Additional ceiling height has cost implications in multi-storey structures, and the equation between ceiling height, cost and the energy savings implicit in the overall provision of daylight, needs to be assessed for whilst the structure is a once and for all cost, the savings in energy which can be obtained continue during the life of the building.

GLASS

Windows or 'wind-eyes' were originally conceived as openings in the building structure through which both light and air was admitted. When these were first filled with glass, it was with the small panes that the manufacturing process of the day permitted, secured by lead beading or leaded lights. As windows developed, larger panes were possible, but little progress was made in solving the environmental problems posed until the latter part of the twentieth century.

There are now a number of different glazing systems designed both to transmit light and to assist in controlling sunlight. The principal purpose

Figure 4.22
One of the galleries at the National Gallery in Washington designed by I.M. Pei and lit from a glass roof. The high level of daylighting makes this an ideal location for the display of sculpture, but makes it unsuitable for the display of paintings. (Copyright David Loe)

Figure 4.23
The side windows to the altar at Basil Spence's Cathedral in Coventry are concealed from the congregation, but provide daylight to the chancel. (Photographer Derek Phillips)

of a window is to provide light and view and the physical properties of all glazing systems should always be studied with this in mind.

Double glazing, followed later by triple glazing, was developed to solve the thermal and acoustic problems of heat loss and noise, but these innovations had comparatively little effect on the entry of solar gain from sunlight. Now new types of glazing systems are available with different properties designed to use new techniques and the architect needs to choose a system suitable for the location and particular circumstances of the particular building.

Glazing systems can be divided into different categories. Here these are divided up in terms of their daylighting qualities rather than those of structure. The material characteristics are not emphasized, since all systems must obey structural needs.

1 Systems which are dedicated to the admission of daylight, do not distort or diminish the view and are used principally to control temperature and external noise.

Single glazing. This can be thick glass, to aid noise control.

Double glazing. Early systems failed due to the inadequacy of the seal between the layers of glass, allowing condensation to develop between the panes. Where severe noise problems exist, large gaps between the glass permit acoustic absorption materials to be placed at the reveals. Electrically controlled blinds may be placed between the layers of glass to assist in controlling solar heat gain and glare. (See under shading systems.)

Triple glazing. Similar in appearance to double glazing, but with increased thermal and acoustic qualities, used where special environmental needs demand.

2 Special coatings designed to reduce solar gain into a space, but which reduce light transmittance, and in some cases distort the colour of the view.

The simplest types are those which rely on a glass coating to reflect the sun's rays, and thus control solar gain. They alter the colour appearance of the exterior façade. Popular coatings are dark and provide a sense of privacy to the interior. However, they also distort the colour of the interior

Figure 4.24
A view seen from inside an office, where the clear daylight view is compared with that seen through a modifying tinted glass. (Photographer Derek Phillips)

itself and where a view can be seen through clear glass at the same time, the impression of daylight is diminished. An alternative is a glass which gives the impression of sunlight even on a dull day (see Figure 4.24).

3 Intelligent systems. These are designed to reduce solar gain, but rely on various means of control that have the visual effect of reducing the daylight admission and view out. It should be borne in mind that where electrical controls are used, they may reduce any energy savings made.

Light-activated or photochromic glass, where the transmission of light to the interior is altered by changes in the exterior conditions, due to reception of ultraviolet light.

Heat-activated or thermochromic glass, where changes in the exterior temperature, alter the optical properties of the glass, and thus the daylight admission.

Electrically-controlled or electrochromic glass formed of a series of layers of glass and other elements where the optical properties can be altered by the introduction of an electric current.

4 Shading systems, internal or external.

Simple internal blinds, or those controlled electrically, have an established use. They are less successful in the control of solar gain, since the heat from the ultra violet rays of the sun will have already entered the building, but they have the advantage of simplicity, and they can be controlled by the occupants.

Slatted or venetian blinds can be incorporated between two skins of glass, but there are disadvantages and advantages in this method in terms of long-term maintenance. They are most appropriate for the control of sun or sky glare, but when in use they cut out views to the outside (see Case Study 35 Cranfield University Library).

External shading will have a marked effect upon the appearance of the building, but has a greater effect upon solar gain, since it acts upon the rays of the sun before they have entered the building. External shading gives rise to its own problems, and must be robust in structure and finish, so that it will withstand all external conditions. It also creates problems of long-term maintenance, which need to be addressed.

DESIGN

There are architectural programmes, such as domestic residences, where the architect's experience, observation and knowledge of the subject will be sufficient to ensure that the plan form will be developed in association with the window design. The aim is to provide interiors well-modelled by daylight and with necessary functional light in order to achieve the impression of the well daylit room.

Daylight design, however, is a specialist subject and architects would be well-advised to arrange for studies to be carried out by an independent lighting consultant or research laboratory, where complex building plans are being investigated. It would be inappropriate to seek advice from a lighting fittings manufacturer, whose legitimate interest is in the design and manufacture of electric lighting equipment.

The architect should always be in control, working to satisfy the aims of the daylighting design together with the need to conserve energy. The daytime use and management of artificial lighting will also have a large part to play in any integrated design in those buildings were daylight alone is insufficient.

Since daylight varies in intensity and direction from morning to night and season to season no finite level can be established. For this reason, the ratio of the light level inside at different points is related to the level of light outside, and this is known as the Daylight Factor (DF). The DF is expressed as a percentage of the light available outside. In overcast sky conditions (5000 lux) a two per cent DF would provide 100 lux.

Using figures for the average level of light available outside in different climates, the DF can be recommended for different types of interior, whilst it is recognized that the level of light at any point with the same DF will vary according to the brightness of the sky outside at any particular moment.

Tables of recommended DFs are available, for different architectural programmes, and if these are followed the amount of daylight will generally be sufficient throughout the day.

Whilst the calculation of daylight inside lit rooms presents little difficulty to the lighting consultant, the calculation of overhead light in atria in multilevel structures and overhead light in buildings with complex sections presents greater difficulties. It may well be best to use the time-honoured system of models, since light reacts in much the same way at model scale as at full size and light measurement can confidently be made at model scale.

Such models should be of a size to allow visual inspection from inside, and be of flexible construction, permitting an investigation of different sectional designs. By taking the model outdoors or by using an artificial sky, a direct comparison can be made between the level of light outside and that within by using photocells to establish the DF at different points in the plan.

However, a model study can do much more than this, as by allowing a visual investigation of the interior spaces modelled by light, it can assist in assessing different window configurations, colour schemes and even furniture layout (see Figure 4.25).

Where the site on which the building is to be placed is obstructed by other buildings it will be necessary to add skyline profiles, to obtain realistic figures which take these other buildings into account.

To gain the interior effect of changes in the sun path throughout the year, the model should be placed under an accurately-positioned sun in an artificial sky, and studies made which can assist in the design of any necessary shading of solar gain.

This is not to denigrate the use of calculations, which together with computer studies can produce a numerical answer, but they still have to be interpreted visually. A model will give the architect a much greater visual impression of complex spaces and will probably take much less time and expense. Daylighting design as we know it today was unknown until the twentieth century, yet beautifully daylit buildings have been known throughout history. The need for new design techniques was made necessary because of the growth of the work place and the economics of lower ceilings in multi-storey buildings.

STRATEGY

It is useful to list the factors an architect must consider when making a decision on the strategy to be adopted for daylighting. Whilst these have

Figure 4.25
Three views of a model of an exhibition area used for the design of galleries at the Tate. (Copyright David Loe)

been put into a logical sequence, it is important to recognize that design is not a linear process. As form follows function, the appearance of a building, of which the window design is such an important element, should be allowed to develop in line with all the criteria for the design. Decisions made initially may need to be reassessed in line with decisions made at a later stage in the design development, a reiterative process.

1 Climate orientation and site characteristics

The basic considerations of where to build related to the climate, and possible orientation may not be in the control of the architect and may have already been made, but the optimum use of daylight should be taken into account at the earliest possible design stage to ensure that there is sufficient daylight entering the building and that it is not blocked by neighbouring obstructions. The course of the sun path should be assessed in terms of the necessity to use and control sunlight. Different locations demand different solutions to the question of solar gain, which may impact on the need for structural shading.

2 Size and proportion of window

Calculate the necessary size and proportion of a window to give an adequate Daylight Factor (DF). An empirical judgement suggests that the amount of glazing should be in the region of 40 per cent of the perimeter area. Overglazing may be a problem. Consider the use of overhead glazing, where the plan demands deep space. Consider the need for an atrium.

To establish the area of glazing required to provide a predominantly daylit appearance it is necessary first to establish by calculation or rule of thumb a mean level or average DF. (There are many guides available to assist in the calculation of Daylight Factor and Average Daylight Factor, and it is not intended to get involved in calculation techniques here.)

The Average Daylight Factor gives a measure of the overall level of daylight in a room. With a 5 per cent average DF the room will have a well daylit appearance, whilst a 2 per cent average DF may require supplementary artificial light in work spaces for much of the time. However a 2 per cent average DF is very adequate in a domestic situation.

It is useful to draw a section through the space and relate this to any external obstructions. A line drawn down from the top of the obstruction through the head of the window will hit the floor at the 'no-sky line,' the point where the obstruction cuts out any direct light, indicating where artificial lighting may be needed to supplement the daylight (see Figure 4.26). Alternatively the answer may lie in some form of bi-lateral daylighting. This decision relates to the need for daytime artificial lighting, and overall energy use. It is clear that the optimum use of daylight leads to energy savings, and with the continuous improvement in the energy efficiency of artificial light sources, further savings can be expected.

Electric lighting should respond to the exterior daylight and to the occupants. Decisions regarding the size and proportion of windows are

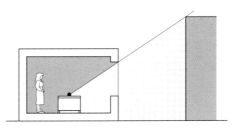

Figure 4.26
No-sky line. This diagram illustrates the demarcation line within a building where due to external obstruction, no view of the sky is visible. This gives a good indication of whether the room will receive sufficient sky light to ensure that the overall impression will be one of daylighting during the day. (James Bell and William Burt, *Designing Buildings for Daylight*, CRC Ltd, 1995)

Innovative daylight systems – Light shelves

(a)

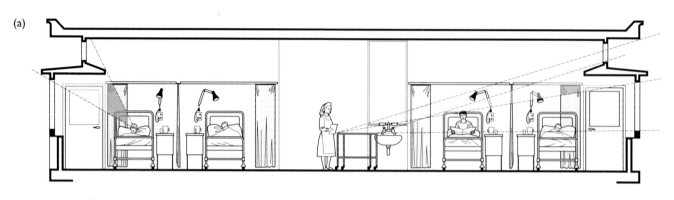

(b)

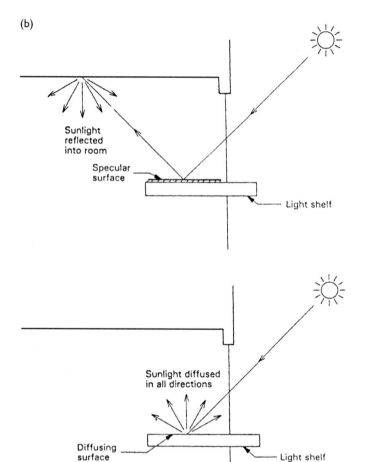

Figure 4.27

4.27 a   An early light shelf at Larkfield Hospital. Reproduced with permission of BRE. BRE publications are available from CRC Ltd, 151 Rosebery Avenue, London, EC1R 4GB. From BRE Report '*Designing with innovative daylighting*' by P.J. Littlefair, published by CRC, Garston, 1996

4.27 b   Reflection from the top surface of two light shelves

4.27 c   Beam sunlighting in a side-lit room. Reproduced with permission of BRE. BRE publications are available from CRC Ltd, 151 Rosebery Avenue, London, EC1r 4GB. From BRE Information Paper IP22/89 '*Innovative daylighting systems*' by P.J. Littlefair, published by CRC, Garston, 1989

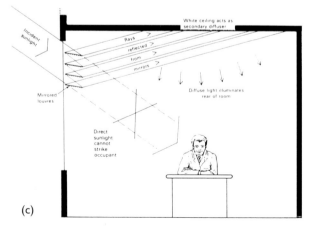

(c)

Innovative daylight systems – Piped light

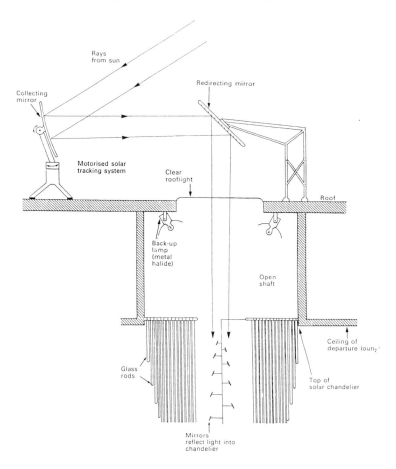

4.27 d   Piped sunlight system at Manchester Airport. Reproduced with permission of BRE. BRE publications are available from CRC Ltd, 151 Rosebery Avenue, London, EC1R 4GB. From BRE Report '*Designing with innovative daylighting*' by P. J. Littlefair, published by CRC, Garston, 1996

4.27 e   Sunpipe   installation   (Copyright Monodraught Ltd).

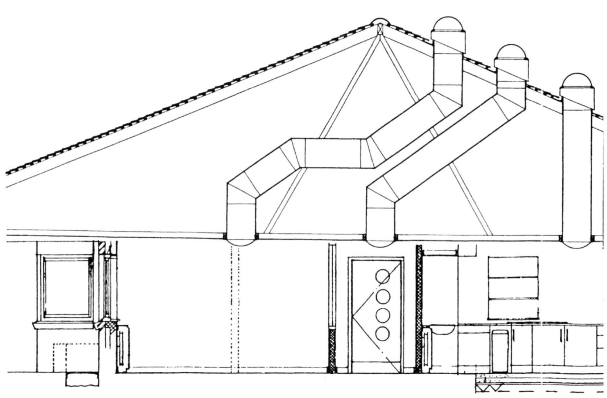

clearly at the heart of daylighting and early preconceptions need to be constantly reassessed.

3 Distribution of light

Consider the distribution of light. Tall windows allow light further into the space. Windows on adjacent walls can assist in providing light to corner rooms. Light distribution is critical for a satisfactory interior environment, and it is therefore crucial to consider the aesthetic requirements of modelling the interior. It is insufficient to achieve enough light (in terms of Daylight Factor) if the resulting visual experience does not give the impression of a well daylit room.

Model studies will generally repay the time spent making them since they can be helpful in assessing not only the levels of light but the visual impact of different proportions of window, together with their light distribution.

4 The glazing system

Types of glazing system have been identified earlier, each having characteristic qualities. It is important to take into account not only the visual environment created within a space, but the impact the glazing system will have on neighbouring buildings, as with reflective glass. Any glazing system will have some structural implications, especially in the case of innovative daylight solutions designed to redistribute light, or special methods of shading to solve the problems of solar gain.

In special situations, innovative daylight systems such as light shelves, prismatic glazing, or holographic louvres that track the sun, might be considered. New methods of gaining daylight within a space are continually being developed.

Daylight can be piped or directed from glazed openings in a roof by a system of distribution similar to the fibre optics of remote source lighting using artificial sources, but in this case using natural daylight (see Figure 4.27).

5 View

The importance of view has been emphasized earlier. The view can be thought of in three layers: the upper, middle and foreground.

The upper layer will generally be the sky, which takes up more of the view the higher you are in a building; the middle is generally the most important part of the view filled with trees and buildings and containing movement; and the foreground may consist of the ground near to the building at low levels.

The most satisfactory views are those containing all three elements or stratas, more easily seen from positions close to the window. If a view is known to be available, dissatisfaction follows when this view cannot be seen.

6 Natural ventilation

Consider the needs of ventilation and whether this can be handled at the perimeter, either by opening windows or by mechanical methods applied to the spandrel detail at the cill level. Ventilation opens up the whole question of the need for air conditioning. It should be possible to make a comparison between full air conditioning, the use of natural ventilation alone or a combination of both natural and mechanical systems. The aspects of window design, thermal mass and passive control of solar heat gain should be assessed interactively.

A further important consideration here is the planned life of the building and how much flexibility should be inbuilt into the design. Experience shows, at least in Britain, that buildings originally designed for limited life are often around for a great deal longer.

7   Control
Finally consider control both of daylighting and of electric light sources. Fine control can save up to 30 to 40 per cent of lighting energy when related to the exterior condition. Consider the possibility of occupant control as there is a basic human need for individual control over one's environment and whilst this may not always be feasible, it should be an aim.

CONCLUSION

Natural light in a building provides variety and interest rarely achieved in any other way and whilst the other advantages of savings in energy and reductions in casual heat gain in the summer are important, environmental quality cannot be over-emphasized. This is consistent with a low-energy policy, where the window design is of the greatest importance.

5 Light sources other than daylight

HISTORY

Up to the end of the nineteenth century artificial light, or light derived from sources other than daylight, was associated with heat, for whilst Sir Humphry Davy had demonstrated the 'arc lamp' as early as 1810, its use as a practical light source had been restricted to providing large amounts of light, as for example in lighthouses. It was not until Swan and Edison developed the carbon filament lamp, or the light bulb as it became known (using a tungsten filament), towards the end of the century that light from electricity became a practical reality.

Initially heat from fire provided light for everyday use. This started some 15 000 years ago with primitive oil lamps made of hollowed-out stone, or clay pots with a wick set in fish or other oil, which, when lit, would have provided a light not dissimilar to that from a candle. These pan lamps are a far cry from the sophisticated oil lamps developed in the nineteenth century where the oils used and the form of lamp provided a brighter and more easily-controlled light.

Another early form of light was the rush lamp, where rushes were dipped in melted fat and lit and in some cases carried around in peoples' mouths called 'splints,' which led to the development of the candle.

The church was the main user of candles made from wax which provided a steady flame, whilst the poor had to make do with tallow candles which guttered and smelt. The candle led to the enrichment of the candle holder, and the quality of candlelight associated with crystal glass in multiple pendant chandeliers gave rise to some of the most beautiful interiors of the eighteenth century.

Of all the original artificial light sources, the candle is the one that still survives because of its particular animation and colour quality, to the point where special flame candles filled with oil are now available.

The greatest innovation in the development of artificial light sources started in the early nineteenth century with the introduction of town or coal gas (known generally as gaslight). This gradually took over from all other sources for lighting buildings, as although it was still a flame source, it was safer and more reliable than earlier types of lighting, with the means of combustion, the gas, being separated from the light itself.

The development of the 'Welsbach mantle' considerably improved the amount of light from the gas flame, which could now be controlled in intensity and which gave an acceptable colour. It was only with the development of electricity for light sources with all its advantages that after a

Figure 5.1
An early oil lamp that used colza oil, a design that has been much copied and converted to electricity. (Copyright Science Museum, London)

Figure 5.2
Early examples of gas lighting for the home and factory. (Copyright Sugg Lighting)

long struggle gaslight was abandoned in the early twentieth century in favour of what we now regard as modern electric lighting.

All the original flame sources identified gave roughly the same sort of flickering light of a similar intensity. A Roman three arm oil pendant of the first century would not have given very different lighting in the room in which it was placed than an early three arm gas pendant eighteen centuries later. Forgetting about the safety and reliability of the light source itself, the chief difference would have been in the quality of the materials used and the form of the pendant itself.

LAMP TYPES

With the introduction of electricity the possibilities open to the architect for designing lighting in buildings to meet the needs of function and decoration entirely changed although initially the early filament lamps were used in much the same way as before. The enormous developments in lamp manufacture over the last sixty years has been exceeded perhaps only by those in computers in meeting the increasing demands of mankind.

Whilst it is not possible in a work of this nature to provide up-to-the-minute data on the many different types of light source now available, it is helpful to identify the main direction of lamp technology and the principle families of light sources and to provide some indication of the advantages and disadvantages of their application.

Incandescent

Tungsten Filament Lamps (GLS)

The earliest filament lamps as developed by Swan in the United Kingdom and Edison in the United States had a short life of only 150 hours and a low efficiency (efficacy) of 2.5 lumens per watt, but they were thought of as a magic light source to replace gaslight. Filament lamps were clean, had greater flexibility, better colour and, with the development of

Figure 5.3
A sketch showing a Roman three-arm pendant illustrates how little has changed in this field apart from the source of power. (Phillips, Derek, *Lighting Historic Buildings*, McGraw Hill, 1997)

Table 5.1
A table designed to show how the various types of electric lamp have developed. Detailed information on the various lamp types is provided in the text.

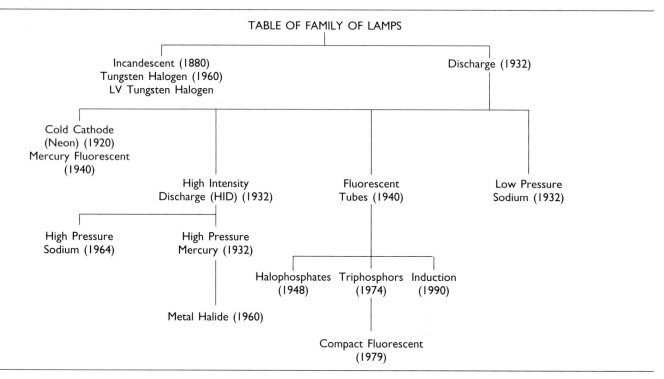

available sources of power, better long-term economics. Now, a century later, the light bulb is still the preferred lamp for domestic use due to its cheap initial cost.

The life of a filament lamp depends upon its light output, with a greater output giving a shorter life. A life of 1000 hours with a light output of 12/14 lumens per watt was established as being a reasonable compromise and one which has stood the test of time. Its colour temperature of 2700 kilowatts gives a warm appearance, which is acceptable domestically, particularly after daylight fades.

The filament lamp had no challenger in the domestic market until recently with the development by 1980 of the compact fluorescent lamp (CFL) with its much improved life and efficiency (efficacy) but with a significantly higher initial lamp cost. The colour quality of the compact fluorescent closely resembles that of the filament and the lamps are small enough to fit with domestic fittings. With the need to reduce the energy used for lighting in the home, these 'compacts' are likely to become increasingly popular (see later).

A significant advantage of tungsten lamps is their capacity to be dimmed, particularly to the designer, who by providing rather more light in a space than might at first be considered necessary can reduce the level whilst at the same time extending the lamp life. They are useful in those areas of a building where it may be desirable to change the light level at different times of day, as in a home or restaurant. Simplicity of control is one of the great advantages of the filament source.

Developments in the field of filament light have taken many forms, but those concerned with the essential characteristics of the source – its ease of control and size – have been the most long-lasting.

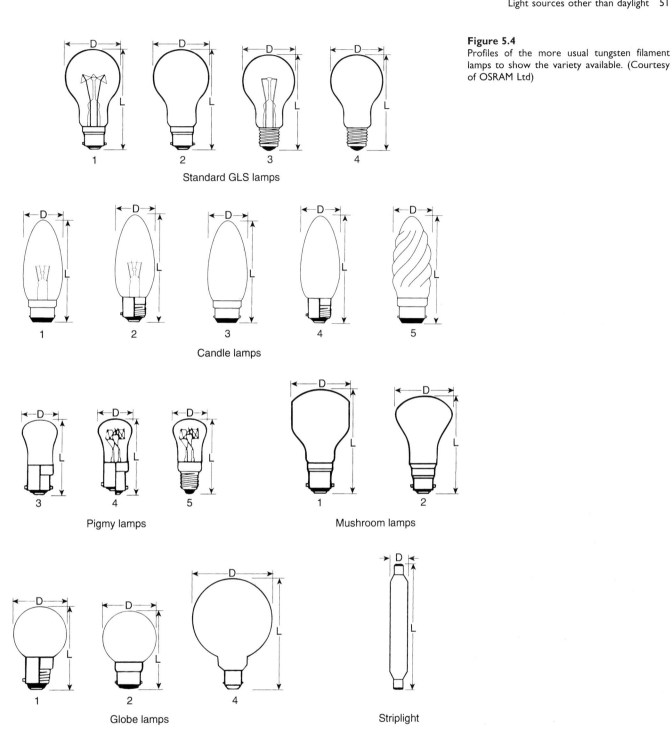

Figure 5.4
Profiles of the more usual tungsten filament lamps to show the variety available. (Courtesy of OSRAM Ltd)

The reflector lamp in its many forms is most useful for controlling the distribution of light. In the case of the crown-silvered lamp all the light is directed back into the reflector to be directed forward in a parallel beam of light, cutting out sideways glare, whilst in the internally silvered reflector lamps, the reflective surface is a part of the lamp, cutting out the need for an external reflector and gathering no dust.

The PAR reflector lamp made from toughened glass has similar characteristics to the reflector, but can be used externally without special weather protection and is often used for simple exterior floodlighting. The advantage of these sealed beam lamps is that the relationship between the

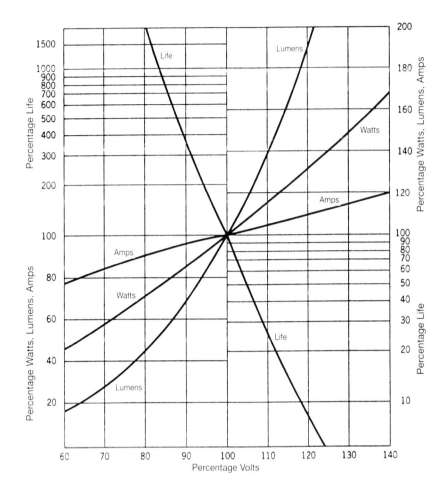

Figure 5.5
This graph illustrates the relationship of the various factors that influence and are influenced by the filament lamp. The most useful is the effect of changes in the supply voltage. For example, a 5 per cent increase in the supply voltage reduces the life of the lamp by 50 per cent; a 5 per cent reduction in the supply voltage increases the lamp life by 200 per cent with some decrease of light output. (Lighting Industry Federation)

filament and its reflector can be accurately controlled to provide a specific light distribution.

The disadvantages of the filament source are its low efficiency (7–14 Lm/watt) and short life, making it an expensive lamp to run for long periods, particularly where access is difficult, and maintenance becomes a problem. It is a high energy source with 90 per cent of the energy wasted as heat in a world where energy is at a premium.

Tungsten halogen lamps (TH)

Tungsten halogen lamps are basically filament lamps with a more robust structure, using a quartz glass envelope containing a halogen gas which allows it to operate at a higher temperature and higher gas pressure. Whilst glass does not transmit ultraviolet light, quartz transmits significant levels which must be considered where halogen lamps are used.

The colour of the light is slightly cooler than normal filament (3000 K), but it is not noticeable, still giving a characteristic 'warm' appearance. Tungsten halogen has double the life of normal filament and better efficiency of up to 22 Lm/watt.

Despite the development of other lamps with a longer life and greater efficiency, tungsten halogen is often chosen for its colour, its simplicity of operation and ease of dimming control. The filament, which is either linear or compact, allows a high degree of control, so that light can be projected from greater distances with greater accuracy, making tungsten halogen a suitable lamp for display purposes, as well as for flood lighting, wall

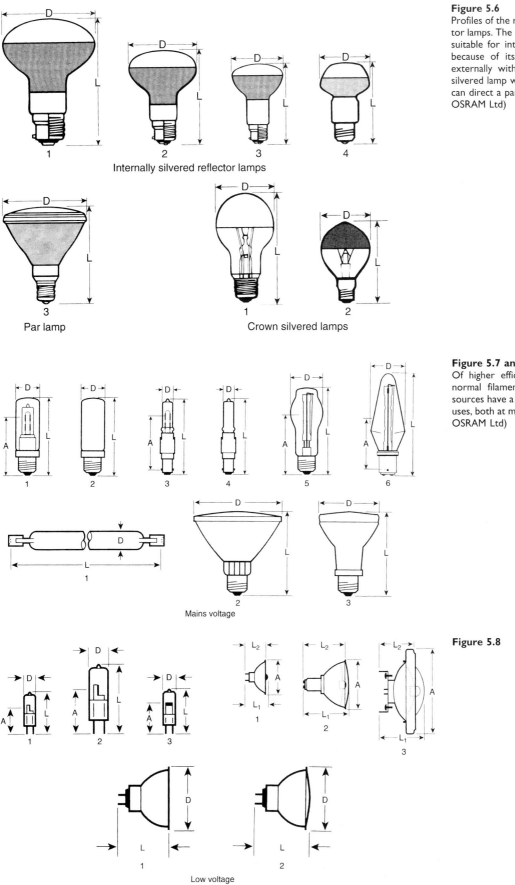

Internally silvered reflector lamps

Par lamp

Crown silvered lamps

Mains voltage

Low voltage

Figure 5.6
Profiles of the range of internally silvered reflector lamps. The most popular is the ISL lamp only suitable for interior use, the PAR lamp which because of its toughened glass can be used externally without protection and the crown silvered lamp which when used with a reflector can direct a parallel beam of light. (Courtesy of OSRAM Ltd)

Figure 5.7 and Figure 5.8
Of higher efficiency and longer life than the normal filament lamps, the tungsten halogen sources have a series of profiles to suit different uses, both at mains or low voltage. (Courtesy of OSRAM Ltd)

Figure 5.8

washing and uplighting. It has a particular relevance to the lighting of historic buildings. (See Phillips, Derek, *Lighting Historic Buildings*, Butterworth-Heinemann/McGraw Hill, 1997.)

Low Voltage lamps (LV)

By reducing the voltage to 12 or 24 volts the filaments in tungsten lamps can be reduced in size, allowing greater accuracy of control. This is a field where there has been considerable development in miniaturization.

Originally developed for car headlights and subsequently for display lighting purposes, low voltage tungsten halogen sources have many advantages, not least in terms of safety. Systems of low voltage display lamps have been designed where the power is supplied by the wires on which they are supported; since contact with the wires at low voltage is relatively harmless provided they are fed from a safety-protected circuit.

The other advantages of standard tungsten halogen lamps also apply the longer lamp life, acceptable colour and simplicity of operation but there is a downside, associated with the provision of power at low voltage. The power needs to be reduced from the mains voltage of 240 volts (110 volts in the United States) down to 12 or 24 volts, which requires a transformer. Such transformers must either be integrated within the design of the individual fitting, on a one for one basis, or alternatively concealed in some suitable position where a single transformer may provide low voltage power to each or several fittings. Problems have arisen with the use of large transformers controlling a number of low voltage lamps, and architects should ensure that when using them, although initially they would seem to be an economic solution, they require an accurately-designed and wired installation.

Discharge lamps

Discharge lamps have no filament; they operate at high or low pressure. All discharge lamps comprise the following elements.

1 Glass envelope
The strength depends upon the pressure required. High pressure lamps have an outer glass jacket for added protection.

2 Electrodes
These are set into the glass envelope, which in the case of fluorescent lamps is in the form of a tube of different diameters.

3 A Gas (mercury or sodium)
When excited by an electrical discharge the gas produces ultraviolet light, with visible light at high pressure.

4 Power
This is used to excite the gas, for example, electricity.

5 Control
This is programmed to limit the flow of energy and start the discharge.

Discharge lamps are very efficient, but some require a length of time to reach maximum intensity, with different types of lamp taking a different amount of time. When some lamps are switched off briefly they will not restrike immediately which can, in some circumstances, present problems.

The nature of the lamp circuitry means that discharge sources are difficult to dim, and even when specialist control equipment is provided for this purpose the lamps cannot be dimmed sequentially from maximum to zero making fade-in effects impossible. They also suffer from flicker.

Colour can also be a problem. With the earlier lamps the problem is in the colour itself and in some later high intensity discharge lamps (HID) the problem is colour shift throughout life. Where colour of light is important (as for example in an art gallery) this is a serious problem. Colour of light, and its constancy throughout life, needs to be assessed when choosing a discharge source in all circumstances. Where conservation of sensitive materials is a concern, the ultraviolet output from discharge sources can also be a problem.

Discharge lamps can be categorized into the following types or families of lamps.

Cold cathode and neon 1920s
Low pressure sodium lamps 1932
High intensity discharge (HID):
    High pressure mercury 1932
    Metal halide 1960
    High pressure sodium 1964
Fluorescent:
    Halophosphates 1948
    Triphosphors1974
    Compact fluorescents 1979
    Induction lamps 1990

There follows some information on each of the lamp families.

Cold cathode and neon

Essentially a type of fluorescent lamp, in that a discharge is created between two electrodes or cathodes, in a tube containing either neon gas or argon with a little mercury which energizes the inner coating of phosphors. The phosphors available allow a wide variety of colours from a range of whites to saturated colours. Neon gas will only provide red light, although all colours are often referred to colloquially as neon light. A variety of tube diameter is available for the lights and these can be bent into any desired shape. The nature of the high voltage discharge creates little heat at the cathodes and due to its construction the power losses are reduced and its life extended to 50 000 hours.

Cold cathode is particularly useful for advertising signs and art works, but in the past the high voltage has meant that special fireman's switches had to be installed to enable the power to be cut off out of reach of the general public. New technology in the form of high frequency gear now permits operation on normal mains voltage, eliminating the need for fireman switches and allowing standard dimming units to be used. The lamps have a very long life, can be dimmed and fitted into situations where access for maintenance may be a problem, or where the ceiling design demands that special contoured lamps need to be bent up, or where there may be special colour requirements.

Low pressure sodium lamp (SOX)

Whilst these are the most energy efficient lamps available, the light is monochromatic, giving a yellow appearance with all colours rendered as shades of yellowish brown or grey. Its main use has been for exterior and street lighting, where its phenomenal output makes its use economic (up to 200 Lm/watt). It is not a suitable lamp for interiors.

High intensity discharge (HID)

High pressure mercury (MBF)

In common with other discharge sources, the high pressure mercury lamp is of moderately high efficiency, and where colour rendition is of little importance this has proved to be a useful lamp. The lamp has been used for street lighting, where its long life and long-term economics have proved valuable, but the lamp gives off light mainly in the purple, green, yellow and ultraviolet wavelengths, with a cold 'blueish' appearance. An improvement is gained with deluxe versions which convert the ultraviolet wavelengths to visible light by means of phosphor coatings. Despite this improvement the lamp is not ideal for interior lighting, except for those programmes, such as warehousing, where the advantages of economics outweigh the importance of colour rendition.

The problem of a long strike time (in the region of five minutes for some lamps) can be overcome to some extent by the addition of a tungsten filament element incorporated in the same envelope, which apart from giving immediate light when switched on acts as the ballast and marginally improves the colour at the red end of the spectrum. The downside of this blended lamp is the decrease of efficiency, to the point where there is little advantage over the tungsten lamp other than its long life. It is little used.

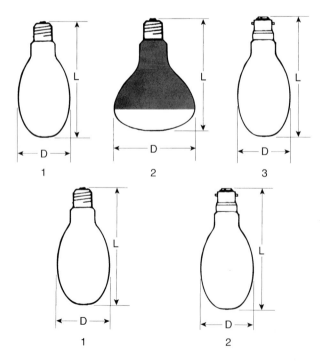

Figure 5.9
Profiles of the high-pressure mercury discharge lamps. (Courtesy of OSRAM Ltd)

Metal halide (HQI)

The metal halide lamp, a development from high pressure mercury, is improved by the addition of halides to the gas which produce a wider spectrum of light and at the same time increase its efficiency. A further advantage of the metal halide lamp is in the wide range of wattage. It can be as low as 35 watts which, because of the small lamp sizes available, allows accurate light distribution of high intensity from small light fittings.

The metal halide lamp would appear to be the answer to many lighting problems: it is efficient, has a long life with a variety of sizes and shapes and is available in warm or cool colour with acceptable colour rendering. However, it suffers from some of the problems associated with high-pressure mercury: it is expensive, requires bulky control gear, has limited dimming facilities with warm-up and restrike periods and does not entirely eliminate the problem of flicker.

Whilst there is a great improvement in the colour of the metal halide lamp when compared to the original high pressure mercury type there is one further problem which has not yet been completely solved. This is the

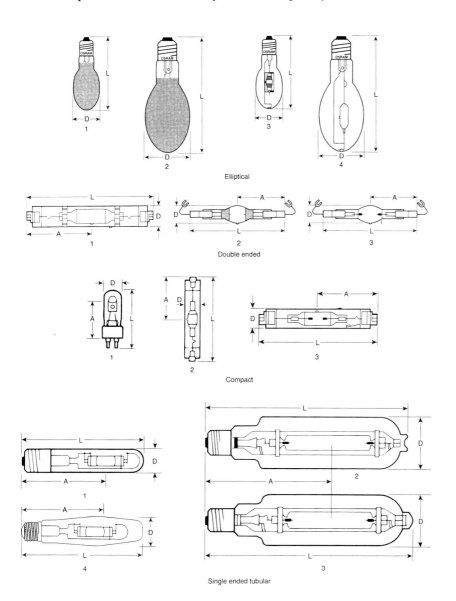

Elliptical

Double ended

Compact

Single ended tubular

Figure 5.10
Profiles of the many types of metal halide lamp developed to be used with a variety of fitting distributions. (Courtesy of OSRAM Ltd)

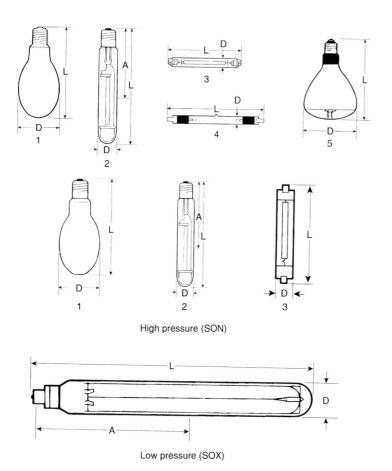

Figure 5.11
Profiles of the low-pressure sodium (SOX) and the high-pressure sodium (SON) lamps. (Courtesy of OSRAM Ltd)

High pressure (SON)

Low pressure (SOX)

colour shift which occurs both in lamps of the same specification and throughout lamp life which makes its use in colour sensitive situations problematic. The technology is being examined by the lamp manufacturers and the problem has been partly solved by new CDM technology.

High pressure sodium (SON)

The higher pressure of the lamp over the original low pressure version improves its colour appearance and the colour rendering of lit objects. The colour rendering is ideal for certain exterior flood lighting projects, the warmth of the lamp colour interacting with the warmth of certain stonework or brickwork, to give good night-time appearance associated with satisfactory economics. The tendency has been to use this lamp as the answer to all exterior lighting projects but it is not; many buildings require a whiter light, as for example Portland stone, flint or concrete, and here other solutions need to be sought.

The colour of the original SON lamp has been improved, and attempts have been made to use it for interior lighting, as for example in airports and other large spaces, without conspicuous success. Having said which, the colour of the deluxe SON lamps is constantly being improved and with future development, its long life and excellent long-term economics, the lamp clearly has a bright future.

Fluorescent, tubular (MCF)

Perhaps the most important innovation in lamp technology, developed shortly before the Second World War, 'the fluorescent lamp' eliminated the heat associated with the original flame sources.

The technology was the same as the low pressure mercury lamps, in that an electric discharge was passed between electrodes placed at opposite ends of a glass tube filled with argon gas, in which there is a small drop of mercury creating a mercury arc; and whilst the electric discharge itself generated little visible light, it was rich in ultraviolet light which acted upon fluorescent coatings on the inside of the tube to provide light in the visible spectrum.

The colour of the light depends upon the nature of the fluorescent coating, and early lamps had poor colour rendering although the lamps themselves were of high efficiency and of comparatively long life (5000–6000 hours).

Fluorescent lamps require control gear, and this coupled with the length of lamp tended to produce bulky and obtrusive fittings which were not liked by architects. However, for economic reasons fluorescent lighting was extensively used.

Halophosphate lamps and triphosphor lamps

The fluorescent lamps developed from 1948 onwards were halophosphates which gave an efficiency of some 72 Lm/watt and they are still widely available. But in 1974 a new range of triphosphor lamps was developed with much improved colour rendering and an efficiency of 88 Lm/watt, which made them a first choice for many commercial installations, where the lamps themselves were generally concealed in recessed fittings. The life of the new fluorescent lamps is rated at 10 000 hours.

The colour from these lamps varies from cool to warm (6500 K to 2700 K) and can be close to the colour rendering of daylight or to that of the filament lamp. The nature of the light generated by the phosphors is not of a continuous spectrum, as it is in the filament lamp, but peaks at certain lines in the blue and green area which leads to colour inconsistency. Fluorescent lamps can be obtained in a variety of diameters and lengths designed to solve problems of integration with other building materials.

From a design point of view the colour rendering of fluorescent lamps is different to the colour appearance of the lamp itself and this can lead to anomalies, where the colour perception of a material seen under different lamps (which may look alike) will appear significantly different. Therefore where colour is of importance, as for example in the choice of furnishing materials, the selection should always be made under the same conditions that will apply in the finished fluorescent lighting installation. Even where a careful choice has been made, the danger that lamps of a different colour may be substituted by maintenance crews cannot be ruled out.

A further drawback of the halophosphate fluorescent lamp is the flicker that occurs at lamp ends and is perceived in peripheral vision. Whilst this has always been apparent (and in other discharge sources), its effects have tended to be played down by manufacturers. It is most apparent where the light is associated with moving machinery and the stroboscopic effect can make a moving part appear stationary. Flicker should not be disregarded as in certain cases it can have health implications.

A significant development in fluorescent lamp technology has been the use of circuits at high frequency (HF). These have many advantages not least of which is that they make it possible to vary the light output from a lamp. The original fluorescent circuits were difficult to dim, whereas the use of HF allows more simple dimming with minimum colour shift between operations within the regulated range. This has great advantages for the designer in many architectural programmes, such as theatrical situations where high and low intensity may be a requirement.

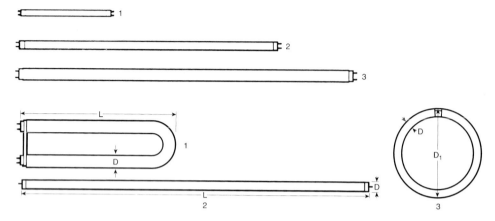

Figure 5.12
Fluorescent lamps come in many lengths, widths and sizes, including U-shapes and circles, this diagram shows a few of these. (Courtesy of OSRAM Ltd)

The advantages of high frequency circuits include higher efficiency and lower energy losses (the use of an HF circuit will increase the efficiency of a trisphosphor lamp from 88 Lm/watt to 96, an energy saving of 20 per cent). A further and not unimportant advantage is the absence of flicker. By raising the cycle from the normal 50 Hertz to 25 000 Hertz flicker becomes imperceptible.

Table 5.2 Proprietary names of fluorescent lamps grouped by colour rendering group (CRG) and correlated temperature (CCT). Colour rendering index values are also cited for each CRG.

CCT (K) CRG: (CRI):	1A (90–100)	1B (80–90)	2 (60–80)	3 (40–60)
Warm (<3300)	Colour 93 Lumilux de Luxe 32 Polylux Deluxe 930	Colour 82 and 83 Energy saver 183 Lumilux 31 and 41 Polylux 827 and 830	Deluxe Warm white	Colour 29 Warm white 29 and 30
Intermediate (3300– 5300)	Chroma 50 Colour 04 Deluxe Natural 36 Lumilux de Luxe 22 Polylux Deluxe 940 and 950	Colour 84 Energy saver 84 Lumilux 21 and 26 Kolor-Rite 38 Polylux 835 and 840	Colour 33 Cool white 20 and 33 Natural 25 Universal white 25	Colour 35 White 23 and 35
Cold (>5300)	Artificial daylight Biolux Colour 95 and 96 Colour matching Lumilux de Luxe 12 Northlight 55	Colour 85 and 86 Lumilux 11 Polylux 860	Daylight	

Reference: CIBSE Code for Interior Lighting p. 96. Published by CIBSE, 1994

Compact fluorescent lamps (CFL)

Despite the wide variety of fluorescent lamp available, the development of the compact fluorescent lamp was a significant step forward as its length and width made it useful in a wider context, albeit with some loss of efficiency. Compact lamps fit into small spaces and their lower wattages make them particularly suitable in situations where filament lamps might have been used in the past.

The triphosphor compact lamps are of good colour and of high efficiency. Two main types are available. In the first type the control gear

Figure 5.13 right
One of the fastest growing families of lamps, the compact fluorescent, or CFL, has expanded considerably. The lamp combines efficiency with good colour, whilst remaining small enough to fit into convenient-sized units. (Courtesy of OSRAM Ltd)

Figure 5.13
Compact fluorescent (CFL).

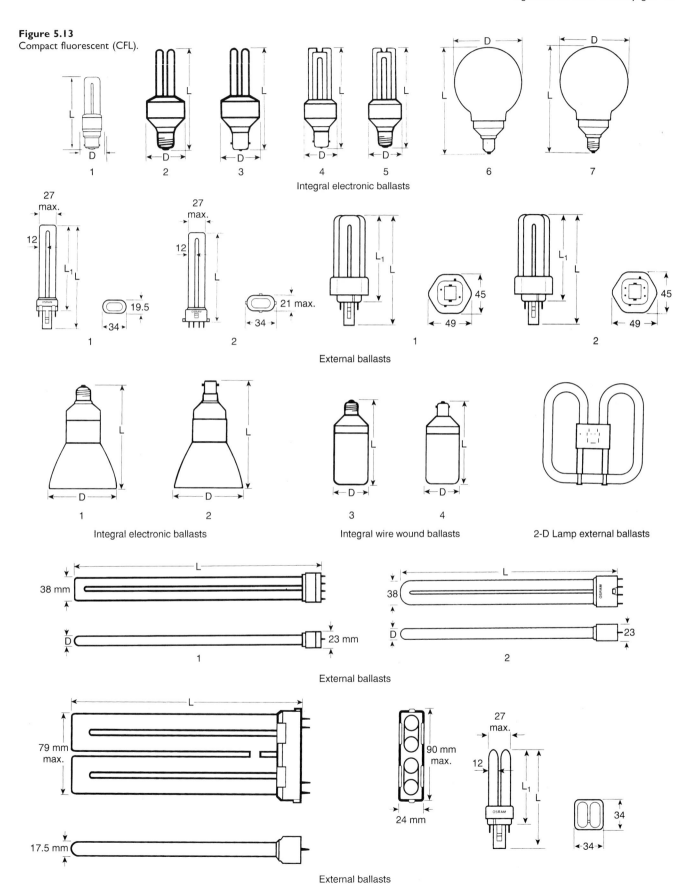

Integral electronic ballasts

External ballasts

Integral electronic ballasts

Integral wire wound ballasts

2-D Lamp external ballasts

External ballasts

External ballasts

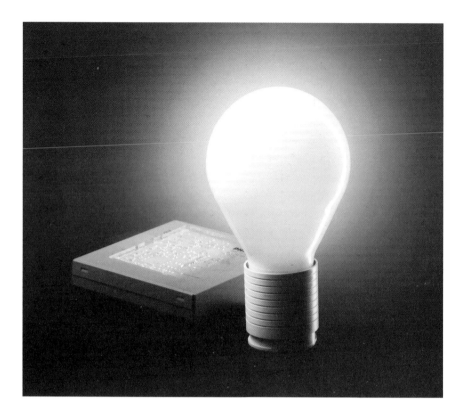

Figure 5.14
Further development is likely in the field of induction lamps which have good colour and extended life once the price can be brought to reasonable levels. (Courtesy of International Lighting Review)

(ballast) is separate from the lamp; in the second, the lamp is integrated with its gear. The second can be incorporated directly into many existing fittings as an energy saving solution, although light fittings designed specially for compact lamps lead to more accurate light distribution.

Induction lamp

The induction lamp although considered under the fluorescent class has no electrodes and operates at radio frequencies. In all other respects it operates on the fluorescent principle: it uses state-of-the-art phosphors for good colour rendering and because there are no electrodes to wear out it has an exceptionally long life (60 000 hours) compared to normal fluorescent lights. Because of its high initial cost, use has been principally for situations where maintenance is difficult or impossible (see Table 5.3, p. 63).

QUALITY

Much research has been devoted to aspects of quality in lighting, but as much of this is subjective, it can mean different things to different people. There are certain aspects which are measurable and the codes of the different Illuminating Engineering Societies make recommendations to this effect.

The level of light for the functional use of the space has been the most greatly researched and the architect cannot be criticized if he ensures that the recommendations of National Codes are met (see CIBSE *Code for Interior Lighting* in the UK).

There will be many architectural programmes where the levels of light are not mandatory and many where they should be variable, leaving the architect to discuss the brief for lighting levels with his client. Art galleries and museums are typical of this, where relatively low levels of light can

Table 5.3 This table illustrates the different aspects of the main types of lamp, providing comparisons to assist the architect in making his choice. The different factors identified are those of efficiency, lamp life and colour, but other factors that must also be considered are those of cost and control.

Lamp	Type	Lamp Efficacy	Circuit Efficacy	Life	Wattages	Colour Temp	CIE Group	CRI
Incandescent	Tungsten Filament	7 to 14 Lm/W	7 to 14 Lm/W	1000 hr	15 to 500 W	2700 K	1A	99
	HV Tung. halogen	16 to 22 Lm/W	16 to 22 Lm/W	2000 hr	40 to 2000 W	2800 to 3100 K	1A	99
	LV Tung. halogen	12 to 24 Lm/W	10 to 23 Lm/W	3000 to 5000 hr	5 to 150 W	2800 to 3100 K	1A	99
Discharge	Cold Cathode	70 Lm/W	60 Lm/W	35 to 50 000 hr	23 W to 40 W per meter	2800 to 5000 K	1A / 2	55 to 65 / 85 to 90
Discharge	Low Pressure Sodium (Sox)	100 to 200 Lm/W	85 to 166 Lm/W	16 000 hr	18 to 180 W	n/a	4	<20
Fluorescent tubes	Halophosphate	32 to 86 Lm/W	13 to 77 Lm/W CCG:48 to 82 Lm/W	10 000 hr	4 to 125 W	3000 to 6500 K	2A to 3	c. 50
	Triphosphor	75 to 104 Lm/W	ECG: 71 to 104 Lm/W	16 000 hr	10 to 70 W	2700 to 6500 K	1A & 1B	85 to 98
Compact fluorescent	Triphosphor	40 to 87 Lm/W	CCG: 25 to 63 Lm/W ECG: 33 to 74 Lm/W	10 000 hr	5 to 55 W	2700 to 5400 K	1B	85
Induction (fluorescent)	Triphosphor	66 to 86 Lm/W	65 to 80 Lm/W	60 000 hr	55 to 150 W	2700 to 6000 K	1B	85
High pressure discharge	High pressure sodium (SON)	75 to 150 Lm/W	60 to 140 Lm/W	26 000 hr	50 to 1000 W	1900 to 2300 K	2B & 4	23 to 60
High pressure discharge	High pressure mercury (MBF)	32 to 60 Lm/W	25 to 56 Lm/W	24 000 hr	50 to 1000 W	3300 to 4200 K	2 & 3	31 to 57
High pressure discharge	Metal halide (HQI)	60 to 120 Lm/W	44 to 115 Lm/W	2000 to 15000 hr	35 to 3500 W	3000 to 6000 K	1A to 2B	60 to 93

1) Lamp efficacy indicates how well the lamp converts electrical power into light. It is always expressed in Lumens per Watt (Lm/W).
2) Circuit efficacy takes into account the power losses of any control gear used to operate the lamps and is also expressed in Lumens per Watt (Lm/W).
3) In the case of reflector lamps where the light output is directional, luminous performance is generally expressed as INTENSITY – the unit of which is the CANDELA (Cd). (1 Candela is an intensity produced by 1 lumen emitting through unit solid angle i.e. steradian)
4) Rated Average Life is the time to which 50% of the lamps in an installation can be expected to have failed. For discharge and fluorescent lamps, the light output declines with burning hours and it is generally more economic to replace lamps before they fail.
5) Colour Temperature is a measure of how 'warm' or 'cold' the light source appears. It is always expressed in KELVIN (K) e.g. Warm white 3000 K, Cool white 4000 K
6) CIE Colour Rendering Groups: 1A (Excellent) to 4 (Poor).
7) Colour Rendering Index (CRI): Scale 0 to 100
where   100 excellent e.g. natural daylight
    85 very good e.g. triphosphor tubes
    50 fair e.g. halophosphate tubes
    20 poor e.g. low pressure sodium lamps

be specified, depending upon the nature of the exhibits and the needs of conservation.

Colour of light is an area in which research continues and where the colour rendition of different lamps is constantly being improved. The colour appearance of the light source and its colour rendition will have an impact on the quality of the installation. Good colour has in the past been associated with higher energy use and higher cost, but now the situation has greatly improved and where quality of colour is important energy efficient solutions are available.

Quality is perhaps the most subjective characteristic that can be applied to lighting solutions; what may provide a certain quality for one person may provide a different quality for another. The dictionary definition of quality refers to a 'degree of excellence', but the word is used here in its more colloquial sense, where the quality may be good, but also poor.

Before attempting to analyse how lighting methods may contribute to it and to try to put quality into some sort of context it is useful to group those adjectives often used to describe the feel of a space.

Comfortable
Pleasant•colourful•light•modelled•restful•quiet•soft•cosy•relaxing•warm•cool (in warm climates)

Bland
Flat•featureless•uniform•monotonous•soporific•boring

Gloomy
Dark•dim•dull•dreary•depressing•oppressive•shady•threatening

Dramatic
Stimulating•theatrical•exciting•bright•interesting•sparkling•glittering•brilliant•varied•concentrated•focussed•intense•glowing•radiant•moody•lively

There are many other adjectives that can be applied to a lighting installation, which may be grouped under one of the above headings – glaring, distracting, obtrusive and dazzling for instance. While a theatrical installation might achieve the effect it is seeking, some parts of it might cause glare from some positions, or be in some way distracting. Similarly a comfortable lighting installation might rely on an unacceptable architectural detail, which would render the lighting scheme obtrusive.

It is worth equating the above list of adjectives with the methods of lighting in normal use:

General diffusing light
Downlights
Uplights
Wall washing
Spotlights
Concealed lighting/remote source
Local light

General diffusing light

Perhaps the earliest form of overall light to a space was given by diffusing fittings designed to give a general light in all directions such as the original Bauhaus spherical glass globes. Inevitably, in order to produce working levels of light, the brightness of the diffusing fitting would have been too high, so that this type was more suited to those areas of a building such as dining rooms or corridors, where high levels of light may be undesirable. Diffusing fittings include the popular noguchi paper shade lamp, many types of chandelier and post top lanterns used at low level in atriums. The popular central ceiling point supplied to so many homes in the early days of artificial light would generally have been equipped with some form of general diffusing fitting which would have been supplemented by table or floor standards.

Downlight

The most efficient form of lighting in functional situations, for example for work, is either mounted directly, recessed, suspended or fixed to track from overhead. If glare is to be avoided, the sideways brightness must be controlled, and if used on its own, light is concentrated downwards on to the horizontal, leaving vertical surfaces dark, with the ceiling unlit.

This leads to a gloomy appearance where the overall impression is depressing. It is usual therefore to use this form of lighting with others to provide some modelling to objects and surfaces. Local light may be used at a person's work position, where it is under his control. Alternatively light may be derived from uplighting, wall washing or spotlighting to provide the necessary degree of contrast and variety.

Uplight

When uplighting is used on its own the ceiling will be the brightest surface in the space. It is an inefficient means of providing a high level of light for work and creates a bland and uniform appearance.

Uplighting is an unsatisfactory lighting system on its own, but it can be associated with other forms of light. It can be useful when low levels of light are required, for example in association with local light sources in domestic residences or other architectural programmes, such as leisure buildings or restaurants.

In spaces with high ceilings, such as historic buildings, where the ceilings are often elaborately decorated and would not otherwise be seen or enjoyed without additional light, a case can be made for uplighting. But even in these circumstances it should rarely be the sole means of light to a space. Other methods should be adopted, allowing the ceiling itself to glow with light rather than light up the space, thus avoiding a bland effect.

The appearance of uplighting in a space is not dissimilar to the illuminated ceiling, popular in the 1960s but which went out of favour for much the same reason as uplights – because it had to be too bright if it was to provide the work light required.

Wall washing

Lighting a vertical wall surface, for example by installing a lighting slot at the top or recessed angled downlights in the ceiling close to the wall, or even by lighting from below, can provide a focus for the eye when used

Figure 5.15
The lighting of the walls to the lifts at the Seagram building in New York in the early 1950s illustrated an excellence of lighting engineering at the time, providing an even light over the full height of the wall. (Phillips, Derek, *Lighting in Architectural Design*, McGraw Hill, 1964, p. 52)

with other forms of lighting to minimize the bland or gloomy appearance associated with downlights or uplights used alone.

The engineering of ceiling slots must be carefully considered. If linear lamps such as fluorescent are used there is always the danger of dark areas from the gaps between the lamps, which may be overcome by overlapping. It is most important that the light across the wall is even and that the lamp colours are consistent, always at risk with fluorescent because of sloppy maintenance.

It is sometimes thought desirable to scallop a wall with light beams by setting ceiling fittings too close to the wall or too far apart. Scalloping is a strong design element in a space and its introduction may be successful particularly where the wall design itself reacts to it, but it should not come as an unwelcome surprise.

Spotlights

Developed originally from theatre lighting, spotlighting shares all the same advantages and disadvantages. Theatre lighting is unidirectional, lighting a stage from concealed positions where glare is eliminated from the eye of the audience, the players themselves being in the spotlight. In the same way, objects surfaces and spaces may be lit using spotlights. The difference is that the direction of the beams of light are not concealed from view as in the theatre and may not be acceptable from all directions in the space so that objects modelled well from one direction become objects of glare from another.

There are some situations where a degree of glare may be permissible, for example in retail display situations, where it is described euphemistically as sparkle, but it is rarely the case and in any situation where comfortable vision is needed glare must be avoided.

Spotlighting is unsuitable for the general lighting of work spaces, but is appropriate for creating positive emphasis by adding vertical modelling to spaces which would otherwise be bland or dull. In museums and art galleries or in situations where an element of theatrical display is required to provide emphasis, spotlighting is an essential design tool.

The miniaturization of spotlighting means it can be concealed in restaurants, for example, where its flexibility will allow changes of table layout.

There is an enormous range of spotlighting equipment available to the designer with almost any beam width and intensity and with both mains and low voltage lamps. The lights are designed for direct mounting or mounting on track and their focus may sometimes be remotely controlled (at a cost). They make a strong design statement but this can be avoided by the designer by placing them in concealed slots, where the impact of the individual fitting becomes a part of the overall building construction.

A disadvantage of spotlighting is its high maintenance. Changing lamps, often at frequent intervals, means firstly that the right lamp is fitted, which is not always easy, and then that the fitting is refocussed to the desired position. In the first enthusiasm of a new and exciting installation this may well be carried out, but in the long term it demands dedication.

Concealed lighting

Concealed lighting as already discussed (see p. 38) is different to uplighting in that it can be applied to any surface in a space. It is rarely used on its own, where it results in a bland appearance, but used with other methods it can add magic to a space.

Local light/task light

In the early days of artificial light, because it was difficult and expensive to provide, light was used locally to meet a person's visual needs at work or play. When electric light became more universally available this was found to be intolerable except in residential situations where pools of light were acceptable. Local light or task light can provide comfortable conditions at work when used with other lighting, but there is always the danger that one person's task light becomes another person's glare source.

Bland or gloomy spaces can be improved by areas of local light, where the light from local sources is directed on to vertical surfaces and used to model planting or other areas of interest. It can be further used to create some change and variation of mood.

Local light allows individual control of one's environment which is an important aspect of man's response to lighting and its greatest use is in residential situations, particularly hotels.

Comfort

To conclude, there is no one definition of a comfortable lighting scheme, since people have different views as to what constitutes comfort. It must be glare-free, but much more than this, it should provide a sense of emotional well-being; it should have a degree of variety, some light and shade; objects should be modelled to reveal their form; it should have a sense of lightness. Comfort is an important element of quality.

The degree of comfort will depend on the nature of the architectural programme and will be different for a library, a church, an office or a squash court and different for a factory, a supermarket or a school classroom. The function of the space will define the elements which make up the appropriate degree of comfort. The lively atmosphere required in a sports hall would be inappropriate in a library reading room, the functional requirements of a garage repair shop unsuitable for a lecture theatre. All these programmes are shown in the Codes as needing the same level of light, but with their very different functions they will have very different needs in comfort.

6 Hardware

LIGHTING METHODS

There are several ways to look at light fittings. Different publications list them with emphasis on the interest of the concern, for example a manufacturer's catalogue will place all downlights together, all track mounted fittings together and so on. Or the fittings can be grouped together according to their light distribution, for example uplight and downlights, but this has the disadvantage that many fittings are designed to provide a variety of light distribution.

The method of lighting, used in the original edition (Phillips, Derek, *Lighting in Architectural Design*, McGraw Hill, 1964) will be examined here as it has the most relevance to architects with their concern with the integration of lighting equipment with the structure of the building. In addition, the design of lighting equipment is in constant development and it is more appropriate to analyse the methods of lighting since these are unlikely to change, whilst the design of lighting equipment inevitably will.

Lighting equipment is divided into the following methods:

Direct-mounted fittings, either ceiling or wall
Suspended fittings
Track-mounted fittings
Concealed lighting/remote sources
Portable fittings

Certain anomalies may occur using this method of analysis of hardware. Concealed lighting, for example, will use fittings that are direct mounted to structure. Furthermore, the desired light distribution will need to be taken into account when selecting the fitting method.

Direct-mounted fittings

The most simple method of support for lighting is where the fittings are mounted directly on to a ceiling or wall structure; where the ceiling or wall must be of sufficient structural stability.

Whilst sufficient support would have gone without saying in the past, the development of suspended ceilings and light partitioning in commercial buildings makes it essential that the necessary support is not neglected.

Ceiling systems have been developed to allow light fittings to be recessed into or mounted on to the ceiling surface, and very sophisticated integrated ceiling designs are available which incorporate all the servicing needs, including air movement, provision for fire and acoustic and climate control.

This method applies to those individual fittings placed in patterns on a ceiling, or in rows or lines, that give a direction to the lighting. Even when service ceilings are not planned, there should be little problem in arranging the electrical distribution in modern buildings where there is space in the ceiling or wall to accommodate wiring, and to incorporate the means for the selected control system to be provided. In older buildings where the structure may consist of masonry walls and the ceilings are formed of primary structure, this can present a problem which is one of the most difficult to solve in many historic buildings.

The structure may not always be a flat surface and may consist of exposed beams and other members supporting the roof above. In such cases the fittings may be attached to these primary members, directed where their location can provide the desired light distribution within the space whilst contributing to its overall appearance.

In some circumstances the structure of the roof itself may suggest a lighting solution where the light fittings are integrated with the structure in such a way as to achieve a synthesis of light and form. An example of this would be where a primary structure offers an opportunity to integrate the light fittings to produce an overall lit appearance, which, whilst providing the building with general light ensures that the whole roof becomes the light fitting.

Clearly here the lighting designer must have an input at an early design stage to assist the structural engineer and architect, to ensure that the needs of the lighting installation are met. The contour of the roof will need to make sure not only that the direct mounted light fittings fulfil the functional needs of the space, but that such factors as the control of glare are addressed (see Case Study 13 Powergen).

Light distribution

It is necessary to differentiate between the light distribution from fittings at the ceiling, which is essentially downward, perhaps with some sideways spill-light to relieve the contrast between the fitting and the ceiling, and those fittings mounted at low level on a column or wall where the distribution can be up, down, or of a more general nature.

Where fittings are fully recessed into the ceiling, often the case in modern offices, the only light reaching the ceiling will be from reflected surfaces, and will be coloured by them (particularly floor surfaces). Where semi-recessed fittings are used with sideways light to the ceiling the light will be at a small angle of incidence, and will emphasize any ceiling irregularity. Ceilings where fittings provide no sideways light tend to give a rather gloomy appearance to a space and other lighting methods may be required to relieve the contrast (see Case Study 10 BA Offices, Waterside (Recessed), Case Study 30 Bisham Abbey Sports Centre (Lines of recessed), Case Study 31 Haileybury School swimming pool (Uplight mounted) and Case Study 34 Eton College Drawing School (Downlight mounted).

Suspended fittings

These are a traditional means of lighting for very good reasons, having easy access for cleaning and maintenance at a low level and having the added advantage of a general distribution of light. The lighting of churches, for example, consists of fittings suspended at low level.

As suspended fittings add an element of furniture to a space they are often of a decorative nature, a typical example being the beautiful glass chandeliers found in historic buildings.

In the twentieth century suspended fittings have been designed that both enhance the appearance of the space through the way the light is distributed and are enjoyed for themselves. When responding to the fashion of the day these have become classic fittings in their turn. For these reasons it is unlikely that where the height of ceiling makes its use appropriate the suspended fitting will ever entirely lose its appeal (see Figure 6.1).

Light distribution

Depending upon the design of a suspended fitting any distribution of light is possible. The most practical is downward with the main light positioned where it is required for functional purposes in the space, but with some upward light to balance the brightness by lighting up to the ceiling (see Case Study 13 Powergen).

Suspended fittings can be used with other forms of lighting to achieve the total design requirement. For example where the downlighting or wall lighting is performed by sources of long life supported at high level the suspended fitting may be required only to add necessary uplighting, itself appearing unlit from below.

With the general lowering of ceiling heights in commercial buildings suspended fittings are rarely used, but they still have a place in architectural programmes demanding a high ceiling level where they offer a very practical solution in terms of efficiency and ease of maintenance.

Track-mounted

The idea for lighting track as we know it is an extension of the original electric busbar trunking used in factories to provide a flexible system of electrical distribution designed to feed lighting and other equipment from overhead lines of electric conduit.

Lighting track was developed in the 1960s, although architects such as Arne Jacobsen had seen advantages in the idea before this. This can be seen in the Rodovre Town Hall where a pattern of spotlights is suspended on electrical conduit, lighting upwards to the ceiling of a conference room. The method of suspension forms an important element of the design.

(a)

(b)

Figure 6.1
(a, b) Whilst it would be interesting to collect together a definitive list of classic light fittings from the beginning of the modern movement in architecture, these two suspended fittings are typical, since they are as functional as when they were designed by Paul Henningson for the Louis Poulson range of fittings and are still decorative today. (Louis Poulson, Denmark)

Figure 6.2
An early use of lighting conduit to support light fittings a forerunner of the many lighting tracks available today (the Rodovre Town Hall, Denmark, designer Arne Jacobsen). (Phillips, Derek, *Lighting in Architectural Design*, McGraw Hill, 1964, p. 130)

At the time of its introduction into the United Kingdom the manufacturer spoke of it as 'solving all lighting problems for any type of installation!' Whilst this was an obvious nonsense, the use of track has had its influence on design and a great variety of track has been developed, from the early single circuit track to be mounted directly on to structure to the multiple circuit tracks. Mains voltage track has been followed by low voltage track, some of which, because low voltage affords little danger, can be expressed as exposed wires to which the fittings are supported, providing a minimal design solution.

The advantages of track systems can be summarized as follows:

1 Ease of electrical installation, since all the light fittings on the length of track can be supplied from a single point. Electrical distribution was and remains the *raison d'être*.
2 With multi-circuit track, different arrangements of light distribution can be provided, on independent switching.
3 Track can be surface-mounted, recessed or suspended.
4 A great variety of light fittings can be accommodated, allowing flexibility of design.
5 Curved track is available, and special contours can be made to suit an architect's requirements. Track can be coloured.
6 Very small section low voltage track is available, to enable miniature spotlights to be accommodated in show cases or for retail display.
7 Gantry systems incorporating track have been designed to carry heavy lighting equipment and can be integrated with other service facilities.

One of the limiting factors of the use of track has been the reluctance of manufacturers to standardize the track so that it can be used with the light fittings of other manufacturers. This is a short term policy and it is hoped that some form of standard can be adopted in the future as at the moment the architect's choice is restricted to those fittings that are compatible with the chosen track.

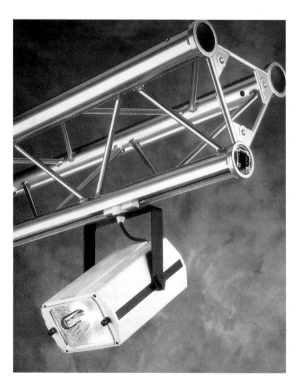

Figure 6.3
A typical gantry system for the support of lighting equipment. (Courtesy of Litestructures)

The supply voltage to a length of track will be constant, so that fittings with special requirements may not be compatible, for example where there is a dimming requirement it will not be possible to use filament and fluorescent on the same track.

Despite all the variety of track available and the advantages outlined above, 'track' is a functional solution which can be obtrusive, and where the track cannot be concealed or integrated into the architect's overall detailing its use is limited.

Light distribution

As with suspended fittings, any type of light distribution can be arranged, depending upon whether the track is ceiling mounted or suspended. Most lighting equipment designed for track mounting is in the form of spotlights, and the rich variety of this type of fitting will ensure that any type of distribution is possible.

Concealed lighting/remote source

In the early days of electric lighting the light fittings available tended to be heavy and obtrusive. Architects disliked yet another design element in a space and for this reason favoured methods of concealed lighting for ceilings or walls. These were described as architectural lighting and whilst eliminating the unsightly hardware they were chiefly associated with inefficiency.

Concealed lighting can be defined as lighting systems where the sources are concealed from view by the fabric of the building and they are generally thought of as a way of providing indirect lighting to walls or ceilings.

The comparatively recent development of remote source lighting can be grouped together with concealed lighting since again the light source itself is concealed by the building fabric; the light being transmitted by fibre optics or other means to the point where the light is required.

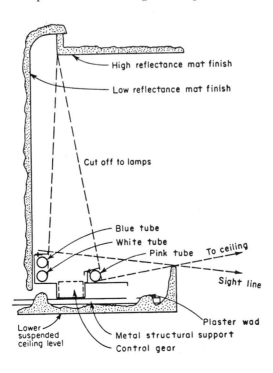

Figure 6.4
A copy of the original cornice lighting detail published in Derek Phillips' *Lighting in Architectural Design* in 1964. The main parameters are still valid today, but more modern light sources should be incorporated. (Phillips, Derek, *Lighting in Architectural Design*, McGraw Hill, 1964, p. 135)

High reflectance mat finish

Low reflectance mat finish

Cut off to lamps

Blue tube
White tube
Pink tube   To ceiling

Sight line

Plaster wad

Lower suspended ceiling level

Metal structural support
Control gear

When using concealed lighting it is important to consider the paramount need to integrate the lighting design with the building structure as early decisions are required if opportunities are not to be wasted and additional expense avoided.

Since the systems rely on the reflective capacity of those surfaces to which the concealed sources give light, there is an important relationship between the light source itself and the colour, texture and reflective capacity of the surface to be lit. The dimensional relationship between the lamps, the surface to be lit and the building structure is crucial. If this is not right there will be too great a contrast of light level across the surface and it will appear unevenly lit. An example of this might be in the design requirement to light a ceiling from around its edge, where the light source must be placed far enough below the ceiling (see Figure 6.4).

There is a danger in thinking that concealed sources on their own can provide satisfactory light to a space. But if, for example, an indirect cornice light is applied to a ceiling the brightest surface in the room will be the ceiling and unless other sources are used the interior appearance will appear bland.

Wall lighting can be successful using concealed slots at the wall to light down the wall, but again the dimensional relationships must be carefully controlled if the wall is not to be bright at the top and dark towards the bottom.

Some of the most successful concealed lighting is baroque in essence, where, as in the churches of Southern Germany the apse of a church is lit from concealed daylight, concealed that is from the direct views of the congregation. This is a technique now used in modern churches, where daylight is combined with concealed electric light for the same purpose. This technique is not dissimilar to theatre lighting and that in many other architectural programmes (see Case Study 9 Fitzwilliam College Chapel).

The great advantage of fibre optics, or as it is now more generally called, 'remote source lighting', is that the light from a single high intensity source can be placed at a distance and concealed where it can be maintained easily. Using flexible glass fibres the light can be distributed to power a number of light sources where they are required and where maintenance might prove to be a problem; only a single lamp will require replacement.

Remote source lighting is essentially inefficient since there are light losses in the transmission of light down the fibres, but this method of concealed lighting has it uses in display lighting and in providing light of a decorative nature. This is of particular use where all the lamp heat and that of associated control gear is kept away from the lit object or any heat sensitive material.

Light distribution

The light from concealed methods is essentially indirect since the light sources are concealed, designed to light other surfaces. With remote source lighting this is not the case, for although the light source itself is concealed, the outlet for the light can be directed to perform functional light where it is required in any direction.

Portable fittings

The first truly portable light was the candle, followed by the oil lamp, both of which could be taken from place to place in a house to give light where it was desired. The favourite expression 'to light one to bed' comes from this time when you were given a candle to see your way upstairs in the days before electric light could be obtained at the touch of a switch.

Attempts were made to provide a degree of portability with early gas lamps using a flexible gas pipe, but the range of movement was limited. It was not until the development of the electric lamp, connected by a wire to a plug, that any real degree of portability was achieved. Even this has the disadvantage that the need for a power supply requires a wandering lead which can prove to be a hazard.

Early portable electric lamps were used for residential lighting to provide local light at low level. These were in the form of floor standard or table light, many of which were very decorative to suit the interior design for the space. Such lamps still have a place in residential lighting in homes or hotels where the accent is on the pool of light illuminating a sitting area for functional or decorative reasons. Where possible a local floor socket should be provided to reduce the length of the supply lead, which requires planning forethought.

The importance of this form of local light is not only one of function; there are psychological advantages in achieving and controlling your own lit environment. Portable lamps provide local light at the control of the individual whilst increasing the amount of light on the work for function (see Case Study 32 Hilton Hotel).

Light distribution

Portable lamps can provide any type of light distribution: downwards, sideways, semi-direct or uplight. A number of modern floor standards provide uplight to the ceiling, reducing the contrast between downward light and what might otherwise be a dark ceiling.

There are specialist uses for the portable lamp in machinery inspection, although this is not normally in the realm of the architect's work. Here preplanning is needed to provide sockets where portable lamps can be plugged in at convenient points to allow machinery maintenance where light will not normally be required.

Whilst 'portables' appear to be a very old-fashioned means of lighting, they are important in creating an individual atmosphere and the fact that manufacturers continue to make large quantities indicates that there is strong demand for them which will continue into the future.

CONTROL

When talking of controls in connection with lighting hardware, it is useful first to make a distinction between the controls needed for certain light sources without which the light source would not operate, known as control gear, and the control of whole lighting systems within a building.

Control gear

Discharge lamps require control gear for them to operate and whilst the control gear for fluorescent lamps is comparatively simple, control gear for the high wattage discharge sources such as metal halide is expensive and it is difficult, if not impossible, to provide imperceptible dimming. Each lamp will require control gear.

Development of high frequency fluorescent lamps and control gear has greatly improved the capability for smooth dimming from maximum to zero, essential where integrated systems of dimming control are required.

Control systems

Systems have been developed to control a whole array of light fittings for various reasons, not least of which is for daylight linking where the artificial lighting in areas of a building are controlled to react to the level of daylight outside, reducing the use of electric energy when daylight is sufficient.

When planning a lighting control system it is essential to understand the nature of all the light sources that are to become a part of the system since it would be impossible to place a system of filament lamps on the same control circuit as a system of fluorescent or discharge sources. The sources must be compatible if placed on the same circuits.

Control systems are at the heart of both energy savings and achieving a variety of effects in an interior. Specialist companies have been developed to provide the necessary hardware to meet the increasing needs in this field.

Control of artificial lighting in its simplest form on/off from local switches – may be all that is required although even in the home the use of dimmer switches have their advantages because they can be used to achieve a variety of effects to suit the individual needs of a room at different times.

A more sophisticated method of control is to link the circuits used in a space together by means of a scene set system. These vary from the simple which might be applied to the home to the highly sophisticated for use in large multi-use spaces such as a cathedral. These pre-sets allow the lighting to change imperceptibly from one scene to another to meet the functional use of the space, either manually by button or by automatic means related to time or to the exterior daylight level.

Control systems associated with the interface between daylight and artificial light should be designed with energy savings in mind. The choice of control system is important and savings of energy of 30 to 40 per cent may be obtained where the daylight and artificial lighting is properly integrated.

In working environments the artificial light can be controlled by photocell or time clock operation to come on when the light from daylight is inadequate and requires supplementing. There are various sophisticated systems that do this. These include lighting circuits linked to photocell operation, presence- or occupancy-operated controls and intelligent luminaires.

When photocell automatic operation is used for daylight linking there is the danger that rapid changes in the sky and cloud formation may result in frequent and irritating changes in the light level from the artificial sources. Means must be available to avoid this, to iron out the changes to an acceptable frequency.

The human need for some control of one's own work environment, for example in matters of heating, ventilation and lighting has already been acknowledged. But it is recognized that in spaces where a large number of people are working at the same sort of task, individual control of lighting presents problems. In such cases it may be necessary to install local lighting at the work station which is at the individual's control. This becomes less necessary where personal computers are used as they provide their own lit ambience. Only where these are not in operation may additional local light be required. There is some danger of over-sophistication, and those systems relying on simple control of lighting circuits may be found to have long term advantages.

The means are available for the use of hand-held infra-red controllers to control individual fittings; alternatively special telephone networks can be employed for the same purpose.

In large building complexes all the environmental services may be controlled by computer using a Building Energy Management System (BEMS). The BEMS in a shopping centre, for example, will make decisions as to the artificial lighting required at different times of day related to the various functions as they change from early morning to night, and to the different exterior lighting conditions at different times of year. Here energy management should again be at the heart of any solution.

7 Building structure

STRATEGY

It is vital that the design for the lighting, both natural and artificial, relates to and informs the structure to provide integrity and clarity.

The strategy for early twentieth century structures was developed by the imperative of daylighting, the structure itself and its configuration which was designed to serve the function of the space. This was particularly evident in the roof forms developed for our factories, designed to maximize the entry of natural light and minimize the adverse effects of sunlight and sky glare.

At the time such structures were designed in the 1940s and 1950s artificial light was both expensive and crude and little was done to integrate the lighting with structure; it was an add-on and often conflicted with the overall integrity of the design. Light fittings and other services hung below the ceiling, eliminating much of the benefit of the daylight.

With the intense development of artificial lighting (both sources and fittings) in the 1960s and its greater efficiency, relative costs were reduced and design turned to structures where the need for a daylight strategy seemed less necessary. This resulted in some buildings where daylight was excluded altogether or used only for environmental purposes, with the main functional light relying on a totally artificial lighting strategy. There were even those who felt that the building could be heated by its lighting system. It appeared that the qualities inherent in daylit structures may have been forgotten.

So much so, that in 1974 the President of the United Kingdom's Illuminating Engineering Society (IES, now the CIBSE) warned of the 'danger that artificial sources will be seen as the only criteria which need be considered, daylight no longer being necessary' and called for a closely 'linked approach where structures should be developed for the introduction of daylight, with an integrated artificial lighting design strategy in which consideration be given to the different qualities of natural and artificial sources.'[1]

The linked approach to structure is now incorporated into most modern buildings, and as shall be seen in the Case Studies in Part 2, the need to develop structures for the admission of natural light with an associated artificial lighting pattern is now well-established.

---

[1] Phillips, Derek, 'Space, Time and Light in Architecture,' IES presidential address, meeting of the Illuminating Engineering Society at the Royal Institution London, 10 October 1974.

STRUCTURE AND LIGHT

A symbiotic relationship between light and structure exists in that the unity of the building derives from their successful interrelationship. In the first instance our perception of structure is gained from the light which falls upon its surfaces and edges: we can only experience structure and consequently the spaces within and without a building as a result of light. Secondly we see the light itself falling upon the structure, either entering through it, by means of windows, or supported from it if it is artificial.

Lighting, both daylight and artificial, determines our perception of structure. In terms of daylight the relationship is unambiguous, since the appearance of a space during the day is generally dependent upon the way in which light from outside enters the interior.

In multi-story spaces this will be through windows on the perimeter, whilst in single-storey spaces this may be through various forms of roof light in addition to or as an alternative to side lighting. Another possibility is the use of atrium design which allows daylight penetration into multi-storey spaces.

A very clear relationship exists between the manner of the daylight and the structural system for the building since an important aspect of the structural strategy will be the environmental criteria of heat, light and sound. Of these light will be a major consideration and the advantages of daylight must be set against the need to overcome its possible disadvantages through structural and environmental solutions.

There are basically two types of structure which influence the nature of the artificial lighting; since they both determine the way in which the equipment is supported and at the same time influence the way in which the space is perceived. These are expressed structure and concealed structure.

Expressed structure

Before the development of sophisticated building services nearly all structure was expressed and whilst its surfaces might have been heavily decorated, basically what you saw was what you got.

This applied not only to the majority of post and lintel structures but also to structures where the roof formations were formed of vaulting or domes. In some circumstances it was considered necessary to form a secondary dome below the external dome (which kept out the weather) in order to provide an acceptable proportion to the interior (see Figure 7.1), but it was not until the needs of other services demanded space to conceal pipes and ducts that the suspended ceiling was developed. Architects took advantage of these suspended ceilings to conceal the wiring and support systems for lighting equipment which resulted in the recessed light fitting.

It is in the nature of expressed structure that there should be nowhere to hide the light fittings and in the early concrete structures of such structural engineers as Nervi and Candela, this caused difficulties, with some unhappy results. It was not until the needs of artificial sources were considered at the design stage that solutions were found.

An early example of an industrial building in the 1950s is that of the Brynmaur rubber factory in Wales (see Case Study 19 Brynmaur rubber factory). This was a wide-span shell concrete construction which permitted large circular openings to be formed in the roof. Half of these identical openings were formed to admit daylight and the other half were designed to provide a home for fluorescent lighting. Roof access allowed

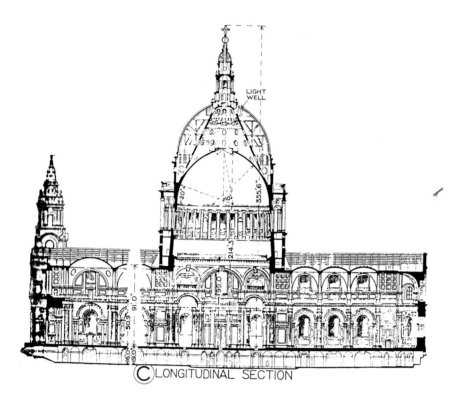

LIGHT WELL

ⒸLONGITUDINAL SECTION

the glazing to the daylight openings to be cleaned, whilst permitting the fluorescent lamps to be serviced and changed from above. This was an early synthesis of both daylight and artificial light with structure.

An example of expressed structure is the cable grid ceiling system employed at the Magnetek facility in Nashville, Tennessee (see Figure 7.2). The system employs both suspended uplights to the ceiling and lines of fluorescent lamps above at the structural ceiling level and is designed to provide the lighting to open plan offices where the layout is flexible. Additional local lights at low level are available to staff who may request them.

The services at ceiling level include air and electrical distribution as well as lighting, and demonstrate the importance which must be attached to determined detailing if an untidy appearance is to be avoided. The use of expressed structure is by no means the soft option; it requires discipline in the planning and integration of the various services.

The advantages of lighting in expressed structures lie in the ease of installation, maintenance and subsequent replacement or adjustment of equipment; the disadvantages lie in the somewhat obtrusive nature of the equipment, which in some circumstances overwhelms the purpose of the space itself. Various gantry systems have been designed to give a homogeneity to the overall appearance of the lighting system, but it is necessary to emphasize that what to one architect is an honest expression of building function may be to another an unwelcome intrusion defeating the primary function of providing light to the space.

Concealed structure

Modern structures demand a high level of servicing, including heating, ventilation, air and electrical distribution, acoustics and means of fire

Figure 7.2
An open office where the structure is expressed, using both direct ceiling-mounted fittings and suspended uplights. (Courtesy of Magnetek Offices, *International Lighting Review*, 1996)

control and as a result space is often provided to house the pipes, ducts and wiring with which these services are associated.

The natural outcome of the revolution in servicing was the design of service ceilings housing the horizontal pipework, using falsework panelling and vertical ducting.

Lighting designers took advantage of these secondary structures to accommodate their equipment, primarily using them to conceal the electrical distribution, then to support the lighting equipment and finally to conceal the equipment by recessing light fittings within the ceiling space. Initially the main structure was used to support the equipment, but this required careful co-ordination with the secondary structure. An obvious development was to make the secondary ceiling the main support for the lighting fittings and to solve the problems of dimensional co-ordination at the design stage of the ceiling and the light fittings together.

To this end suspended ceilings were designed not only to conceal the pipe and ductwork, but to integrate all the services within the ceiling space, including the lighting. Lighting often took the form of individual recessed fittings, or lines of light, the shape or form reflecting the nature of the source, fluorescent suggesting lines, filament and other compact sources suggesting circles.

It is important that the lighting designer co-ordinates his work with that of the structural engineer to ensure that the design concepts of the architect are achieved. It is equally important that the lighting designer co-ordinates his work with that of the services engineer responsible for the pipes and ductwork.

One problem already identified with recessed lights set into a suspended ceiling is that of the brightness contrast between the fitting against the dark, unlit ceiling which causes a gloomy interior. The problem is exacerbated by the need to overcome the problem of glare and the associated problems of working with VDU screens by using low brightness louvre systems.

One solution is to provide upward light to the ceiling area, but this is not always practical. An alternative is to contour the ceiling to the needs of other servicing or structure, leaving central areas in each bay from which can be suspended lines of downward lighting to provide the main night-time lighting for the building.

These lines of downward lighting would on their own suffer from the problem of contrast already identified and special equipment has been developed to provide upward light from the top of these fittings to illuminate the flat ceiling area above. By switching this separately there is an added advantage, that where a daylight strategy has been developed for the building, the downward lighting energy may be reduced (or in some cases eliminated) using the upward light to alleviate the darkness on the ceiling and greatly improve the perception of the space (see Case Study 12 BRE Environmental Building).

The structure as light fitting

In a sense this is the opposite to the concealment of the structure behind secondary elements. Here the structure itself is designed to control the light and itself becomes the light fitting.

There are many examples from history of structures designed to control the admission of daylight, as for example the side bays at Zwiefalten Abbey (see Figure 3.5) which limit the views of the congregation in order

Figure 7.3
The Citroën styling dome at Velizy, lit upwards around its edge from powerful light sources, is virtually on the same principle as the GM Dome some forty years earlier. This is hardly surprising since the purpose of the building is identical and it would have been difficult to engineer the design in any other way. (Courtesy of *International Lighting Review*, 1995)

to eliminate glare whilst allowing in daylight. This is an example of daylight from the side but many of the baroque churches in southern Germany used the fabric of the church to reflect light into the interior whilst concealing the source of daylight.

Whilst this applies more generally to the use of natural light, it can also be applied in some circumstances to the use of artificial sources, for example the styling dome at Citroën, Velizy, where the whole vault of the dome is indirectly lit by artificial lighting from around the edges; the dome itself then might be considered as the light fitting. In this particular case it is performing a specialist function, that of conveying an impression of the daylight outside, to enable an assessment to be made of new models of cars.

Co-ordination of structure with lighting elements

Lighting equipment will generally be associated with forms of building structure whether these are primary elements such as walls, floors and roofs or secondary elements such as suspended ceilings, partitions or built-in furniture.[2]

It is necessary to apply a different approach to each. Light fittings are manufactured to the close tolerances akin to the factory processes of secondary structures, whilst much primary structure is craft-based (as with concrete and brick). In these circumstances it is unwise to rely on a close-fit relationship between the light fitting and its structural support – a loose-fit arrangement allowing for some degree of structural diversity would be more appropriate.

[2] Boud, J. and Phillips, D., 'The relationship of lighting equipment to the fabric, finish and furnishing of buildings,' IES National Conference, York, 1976.

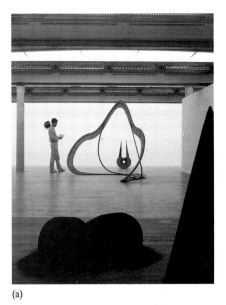

(a)

(b)

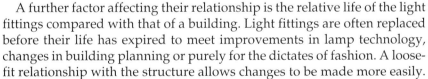

(c)

Figure 7.4
(a–c) These three illustrations show the different appearance of a wall when lit differently: (a) the wall is evenly lit; (b) the wall is scalloped by beams of light; and (c) the wall is not lit at all, deriving sufficient light for its purpose from the corridor lighting. (Courtesy of Zumtobel Lighting)

A further factor affecting their relationship is the relative life of the light fittings compared with that of a building. Light fittings are often replaced before their life has expired to meet improvements in lamp technology, changes in building planning or purely for the dictates of fashion. A loose-fit relationship with the structure allows changes to be made more easily.

In the case of secondary structures, where the fittings have been co-ordinated to become a part of the ceiling, wall or furniture design, it is likely that both will be replaced before the life of the building is ended and a new co-ordinated relationship established with the updated lighting equipment.

Structural integrity and clarity

Light can alter our experience of structure to give it solidity or weight, to fragment it, or indeed make it disappear altogether, and the architect should be aware both of the power of light to modify the impression he wants to create – for this should not be achieved by accident but by mutual understanding between the architect and his lighting designer.

Taking a wall as an example, the following set of photographs illustrate the different ways in which light sources can be related to the surface and some of the variety of effects achieved (see Figure 7.4).

It would be impossible to illustrate all the different structural elements in a building and the way in which light may be used to provide integrity and clarity, nor is this something that can be brought together in a book, rather this is something that the architect learns by experience and most of all by observation. What is certain is that in order to make the right decisions the architect should understand the way in which light can inform or destroy the visual quality of his structure.

(a)

(b)

Figure 7.5
(a, b) A day and night view of the same building, The Manufacturers Trust Building on 5th Avenue, New York, in 1953. The impressions are very different, creating a solid building during the day and a glass box at night, but the building is clearly one and the same. (Photographer Derek Phillips)

NIGHT APPEARANCE

It was in the late 1950s that the architect Gio Ponte writing in the *International Lighting Review* spoke of the 'second aspect of architecture' or the appearance a building takes on after dark. He gave as an example the design for the Pirelli building in Milan, where he had planned the accommodation in such a way as to ensure a satisfactory relationship between the building as seen by daylight and as experienced after dark with its internal lighting.

Up to this time it would have been more likely for the exterior to be lit by some form of floodlighting, as was the case in most historic buildings where solids tended to dominate over the voids in the façade. However with the revolution in construction of the modern movement in architecture and the greater use of exterior glazing it was possible for areas of the

Figure 7.6
The Lloyds of London Building, seen here at night with added blue floodlighting, is a city landmark. (Photographer Derek Phillips)

building, or indeed the whole building, to appear as a glass box lit from the inside at night (see Figure 7.5).

What is important in terms of the structural aspect of the night appearance is that the building should show a correlation with its daytime appearance, for while not being the same (indeed it may be quite the reverse, black being light and light black), it should appear to be same building. It is therefore important to consider the appearance of the structure at night when considering the overall strategy for the plan.

It is not uncommon, for example in art galleries or theatres, to have blank walls on the inside to exclude daylight and to surround these with glass exteriors. Whilst daylight enters the building from the exterior through the surrounding glazing, it is the lit interior walls that are experienced from the outside of the building at night. Since the glazing cannot be lit, the lighting of these walls is of the greatest importance.

The evocative use of colour for the night-time lighting of building structure has been influenced by the greater efficiency of modern light sources. The use of colour for lighting buildings, such as Lloyds in the centre of London, whilst creating a dramatic effect does not alter the correlation between the day and night appearance – it is still manifestly the same building, its appearance after dark having resulted from conscious decision (see Figure 7.6).

Likewise the lighting of the Waste Disposal Plant at Tyseley, where colours change with the time, has given a completely new look to what during the day might seem to be a rather banal structure (see Case Study 17 Waste Disposal Plant, Tyseley).

The architect's relationship with the structural engineer has to be sufficient to provide an overall knowledge of what is possible and the same is true of his knowledge of lighting design: the architect must be aware of the principles so as not to ask the impossible, but also be on the lookout for developments in lighting in order to expand the boundaries of what may be possible.

8 Installation and maintenance

ENERGY

When choosing a lighting system for a building the energy use of the system is of increasing importance, both in terms of its initial capital cost (generally, the lower the capital cost the higher the cost of energy) and in the running cost where the opposite may apply. Low energy prices tend to lead to greater energy use.

Lighting, like other energy sources used in buildings, relies on the production of energy from fossil fuels of which there is only a finite world reserve. In addition, buildings are the largest source of the release of carbon dioxide, contributing to the greenhouse effect and the danger of global warming. There is, therefore, every need to try to limit and, if appropriate, reduce the amount of energy used by lighting in buildings.

The most important way to do this is to use natural light as far as possible by designing the form of the building so as to maximize the use of daylight as the functional building lighting during daylight hours. For example, the proposed building footprint (square or rectangular) will determine the amount of wall area and thermal quality as well as determining the daylight characteristics.

Discussion of the problems of energy use in lighting are dealt with under the following headings: building function; relationship between artificial light and daylight; control of light; colour; and costs, capital and running.

It must be emphasized that the energy expended on lighting is only one of the ways of energy use in buildings; there are others, but lighting is perhaps the most significant use of electricity.

Building function

The function of a building will determine the amount of light required. Some buildings or areas of buildings will demand a high level of light to serve the visual needs of the occupants, whilst other areas may be satisfied by lower levels. Suggested levels are found in publications such as the Codes of Practice of the different Illuminating Engineering Societies (see Chapter 3, p. 10).

Whilst it is essential to satisfy the visual needs of those using the building, it should be done in as energy efficient manner as possible. When considering the energy aspects of the different lighting systems to be evaluated, it should never be forgotten that the chosen system must satisfy both the emotional as well as the visual needs of the occupants – seeing and perception.

There are programmes where the emotional needs of the occupants are at least as important as the visual needs. A good example of this is the lighting of a hotel restaurant where a low-energy system may be compared with one where the total energy consumed is higher. It would be wrong to settle for a low-energy system of poor colour that lacks flexibility of control rather than a system using higher energy but with excellent colour and dimming facility. In the first case the restaurant is likely to be a failure and in the latter it will have a chance of success.

So energy efficiency, whilst being an important criterion, must be judged in terms of how the lighting system meets, or fails to meet, the visual and emotional needs of the space.

Relationship between artificial light and daylight

It has been stated at the outset of this section on energy that the most important contribution to its conservation will be to maximize the use of natural light during the day. The introduction of daylight through windows, where the daylight is designed to contribute a substantial element of the daytime lighting, is a basic design decision affecting the whole design of the building. It will affect its orientation on the site, its means of ventilation and sky and sun glare control and it cannot be thought of as some sort of add-on after the building form has been decided.

At a conference at the RIBA in 1996[1] a figure of 40 per cent glazing of the perimeter area was suggested as leading to significant savings of energy in terms of the artificial lighting, size of plant rooms and improvements in the internal environment, leading on to greater human productivity.

The design of artificial lighting in relation to daylight in buildings has had a long history. It started with totally daylit buildings with tall ceilings and windows that allowed daylight to reach the rear of deep spaces and where the artificial sources of light were used only after dark. When for economic reasons ceiling heights had to be reduced, it was found that insufficient daylight reached the rear of the space and the concept of Permanent Supplementary Artificial Lighting (PSALI) was developed,[2] where daylight was used to supplement the artificial light.

The concept of PSALI in the 1960s had much to do with the expectation of high levels of light in offices as well as with deep plans, where the centres of such spaces received little daylight. Where the centres of deep spaces required artificial light at all times, similar levels should be available close to the windows during the day, and this could rarely be obtained by daylight alone. The *raison d'être* of this concept was more concerned with the quality of the lit spaces than achieving savings in electricity.

Daylight penetration tended to be related to the amount of façade devoted to windows. Due to the environmental design of the 1960s this was often limited to 20 per cent, which made it necessary to combine artificial light with daylight during the day, with little saving in energy.

---

[1] 'New Light on Windows,' RIBA/CIBSE Seminar, November 1996.
[2] PSALI paper by Ralph Hopkinson and James Longmore, 1959.

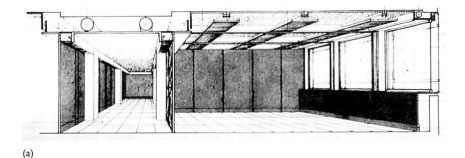

(a)

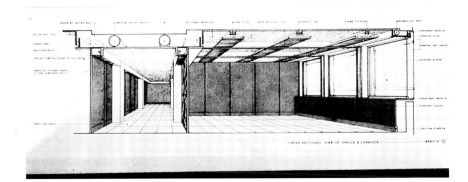

(b)

Figure 8.1
(a, b) Esso Building, London. An early example of Permanent Supplementary Artificial Lighting (PSALI) in 1959. (Architect Sir Denys Lasdun; lighting design DPA Lighting consultants)

In this example a distinction is made between the distribution of light during the day and at night. Each fitting contains three lamps, but during the day the two rows close to the rear wall use all three lamps whilst the row at the window where the daylight is adequate is turned off. (Daytime total energy use = 6 lamps.)

At night all three rows use two lamps to provide an even design level over the whole office. (Night-time total energy use = 6 lamps.)

This is an early example of the control of energy use before the advent of sophisticated controls and daylight linking.

The main change today is the requirement to both enhance the quality of the interior and reduce the need for artificial light during the day, thus saving considerable amounts of energy. Indeed the use of photovoltaic panels on the façade can provide sufficient energy for the artificial lighting in those areas of the building where daylighting is inadequate, while permitting sufficient daylight to enter the building during the day, obviating the need for electric lighting in the perimeter area altogether. This requires an integrated approach to building design (see Case Study 11, Solar office, Doxford).

Control of light

Forms of control are at the heart of energy saving in the lighting of buildings. These are discussed in their various forms in Chapter 6 (Hardware) where it can be seen that control of lighting is now very sophisticated, with systems designed to react to the levels of daylight to ensure that electric light is reduced or not used when it is no longer required.

Control of natural light is concerned with the need to eliminate glare from the sun and the sky by using shading. The admission of natural light through windows of all types must be seen in its context as it affects the amount of heat gain and loss which has implications in terms of ventilation and air movement.

Large building complexes are often seen after dark with all levels of lighting on when it is clearly not being occupied, or is only partially occupied and this is most wasteful of energy. Modern control systems working through the BEMS (Building Energy Management System) are designed to ensure that this does not happen. The cleaning and maintenance of such buildings should be planned to avoid wasted energy of this sort, lighting levels being related to the element of servicing to be carried out, whilst left at security levels in other areas. This is a field where energy savings should not only be possible, they should be mandatory.

Colour

Natural daylight is the colour standard by which artificial light is judged. Daylight has a continuous spectrum and although it differs from early morning until evening, the change experienced is understood and accepted, so that a white wall will still appear to be white despite its physical change. In fluorescent light the colour of perceived light and its effect upon the colour of objects and surfaces in a space results from the relationship of the mercury arc activating the phosphor coating on the inside of the tube. For this reason the colour rendition of fluorescent lamps varies from poor to acceptable. The most efficient lamps in terms of energy have in the past been associated with the poorest colour production, but recent improvements in the colour rendering of the fluorescent triphosophor lamps and the latest metal halide lamps now means good colour is associated with high efficiency, a relationship that will no doubt continue to improve.

The CIE (Commission International de L'Eclairage) divides the colour rendition of lamps into five groups, which are shown here in their relation to the Colour Rendering Index (CRI):

1A	Accurate colour matching	CRI 90–100
1B	Good colour rendering	CRI 80–89
2	Moderate colour rendering	CRI 60–79
3	Colour rendering of little significance	CRI 40–59
4	Colour rendering of no importance	CRI 20–39

Cat. 1A
Apart from tungsten filament and tungsten halogen lamps few others meet this demanding standard, although specialist fluorescent lamps and some metal halide types are now available (particularly for use in colour printing inspection).
Cat. 1B
The range of fluorescent triphosphor lamps fall into this category; they are suitable for commercial and industrial use where colour is of importance. In addition the compact fluorescent lamps, some metal halides and the white SON are included.
Cat. 2
These are used where colour rendering is of less importance, as in some commercial premises. It includes the range of halophosphate fluorescent lamps and the metal halide lamp.
Cat. 3
Many types of lamp are available in this category, where colour rendering is of little significance, to include the high-pressure mercury and sodium lamps.

Cat. 4
Finally, where colour is of little importance the standard high-pressure mercury and low-pressure sodium lamps will be satisfactory.

Table 5.3 (p. 63) shows lamp types associated with both the CIE Colour Rendering Groups and Colour Rendering Index (CRI). When making a choice of lamp, it is essential to ensure that the colour rendering falls within an acceptable category. Having said this, it is still advisable to make the choice of decorative materials under both daylight and the alternative light source.

Costs, capital and running

When planning any lighting installation an analysis will be made of its cost – both its capital and its running cost – so that a cost over life can be compared with alternatives. Decisions as to the long-term and short-term economics should be made after such a study.

Whilst not the whole energy story, this analysis will enable a client to judge the effect of the different parameters of each system, when finalizing the light source and associated equipment. It will then be possible to relate the energy use cost to other factors. There is a danger that this may be distorted where the cost of electricity is cheap, leading to an increased use of energy.

To make this calculation, certain facts must be established:
The level of light (illuminance) required to meet the function;
The implications of any daylight interface;
Whether there will be determined maintenance;
The adequacy of the colour rendition of the lamps to be compared;
What lighting controls should be provided.

1 Capital Cost
(a) The initial cost of the lamps and lighting equipment together with the cost of the system design.
(b) The cost of the electrical installation together with associated controls.
(c) Any additional building costs, due to special structural configuration made necessary by lighting integration.
(d) Interest on capital employed.
Note: estimate the total cost of installation and divide this by the number of years of operation (usually set at ten) to provide an annual capital cost.

2 Running Cost
(a) Use the efficiency of the lamps to be investigated to calculate the annual cost of electricity. This will need to take into account the efficiency of the lighting system to be employed and the number of hours the installation will be in use. For example, an installation of upward lighting will use more electricity than one of more general distribution for the same light levels, and the running cost will be reduced for installations such as churches which have limited daily use.
(b) Calculate the reduction in energy use made possible by the electric light/daylight interface. This can vary from almost nil to a serious reduction in the use of electricity during the day.
(c) Examine the life of the lamp, its cost and its location in relationship to the building structure to calculate the annual maintenance cost of the installation.

The annual capital cost must then be added to the annual running cost to give the total annual costs for the lighting. If this analysis is carried out for a variety of lamps an informed choice of lamp can be made.

However there are a number of factors which should be borne in mind when making comparisons. Daylight integration will have a cost, which needs to be considered. The function of the building will impact on the decision; what might seem to be an economic decision for one class of building might be a disaster for another. The emotional needs of the building should not be disregarded either (refer back to the hotel dining room analogy).

In large building complexes, such as office blocks, shopping centres or hospitals, a system of energy management will be installed. This BEMS (Building Energy Management System) can be programmed to ensure that energy is not wasted, by reducing system light output and energy use when not required, and through an interface with daylight.

On smaller schemes where no BEMS is planned, the question of energy management should not be disregarded, and this may mean the appointment of an energy manager, or a person whose role it is to monitor the use of energy throughout the building to avoid wastage and to seek ways of saving energy as new technology becomes available. In order for him to do this it will be necessary for whoever is responsible for the electrical installation to provide this person with the necessary documentation on the commissioning of the building. A further duty of the energy manager will be to monitor the life of lamps, to ensure that appropriate lamp changes are planned to maintain the required illumination levels.

An aspect which can have an impact on costs in those programmes, such as hotels, where a single element is repeated many times, is the economics of having a mock-up built, where it is possible to judge the materials, colours, and visual quality of the lighting to produce a more competitive tendering.

INSTALLATION

Except on small projects, it will be the task of the electrical consultant to plan the electrical installation, but it will be the architect's role to see that it is done in a cost effective manner. It is important that the electrical installation should be easy to understand and to install, and that it should meet the client's brief as to how much flexibility should be provided to avoid the need for system changes with subsequent accommodation changes.

Where possible, the installation should be planned for ease of access, but where this is difficult, as for example in large, tall spaces, other means may have to be adopted. The lamp must be selected carefully, to ensure that it has as long a life as possible, consistent with the function of the lighting system. With short-life tungsten filament lamps, these may be underrun to increase their life. Overhead access is one solution, as for example in the catwalks above the sports hall at Bisham Abbey (see Case Study 30 Bisham Abbey Sports Centre) where the lighting equipment may be maintained without disturbing the play below.

One of the reasons for the development of the various forms of lighting track was simplicity of installation; multiple circuit track, for example, provides both flexibility and ease of installation, but is not the answer in all cases as it is not always visually acceptable.

The choice of the lamps and the lighting fittings to be installed is clearly crucial when planning the electrical installation and the associated control systems.

MAINTENANCE

Planned maintenance is essential and is dependent upon the method of installation, the choice of lighting equipment, its quality, its lamp and lamp life and the design. What might at first appear to be an inexpensive solution may in the long run be more expensive, where the design has not been fully developed and units are installed that are difficult to clean and maintain.

The increased life of many of the available lamps, particularly in the field of discharge sources, may lead to a reduction in maintenance; and whilst this has cost advantages it should not be at the expense of the gradual loss of light which occurs through dirt collection and the normal diminution of light output from old lamps. A system of bulk lamp replacement when the lamps reach a certain length of life, carried out without disruption at a time when the area of the building is not in use, may have advantages.

Where large pendant fittings are used in tall spaces, it may be economic to arrange for these to be mechanically lowered to the ground level with special raising and lowering gear to enable a maintenance crew to clean and re-lamp. The gear and its support is expensive and will need to be fully investigated at the time of installation.

Safety

The design and installation of the lighting system may be the responsibility of an electrical consultant, but the overall safety of the building is still in the hands of the architect, and he must ensure the following:

1 The adequate safety of the light fittings and electrical circuits against the possibility of electric shock. All lighting equipment in the United Kingdom must pass the relevant British Standards, particularly so where fittings are imported. The operating voltage in the United States is half that of the United Kingdom (110 volts) and for that reason is less dangerous, but the standards of safety of light fittings required in the United States are led by the insurance companies and are stringent.
2 Similarly, the necessary standards for protection against fire of the whole electrical installation.
3 Overall standards of electrical safety during construction on site are observed. Whilst more directly the concern of the building contractor, this is still important and for this reason power tools are generally run at low voltage. As the architect has an overall duty of care while the building is under construction, it is important that during his site visits he or his representative should spot any irregularities.
4 A satisfactory system of emergency lighting is installed and agreed with the fire officer. The greatest danger here may be that of smoke and special arrangements may need to be made to provide escape routes identified at low level. Emergency lighting is not something that should be thought of as an afterthought, but should be the responsibility of the architect and lighting designer and planned as an integrated part of the overall lighting design.

It can be seen from the above that the question of installation and maintenance cannot be ignored by the architect.

9 Building services

INTRODUCTION

When discussing the relationship of lighting to building services it is important to make a distinction between those aspects of lighting which are concerned with seeing and those which relate to the integration of lighting with other service elements.

Level 1: aspects concerned with seeing – the visual experience

For architects this must have the highest priority since it is light that enables a person to see and to perform or function within a building. Light allows a person to experience the spaces within a building and ultimately to enjoy the spaces and their architectural character. This experience is evident as much during the day as at night, and the relationship of daylight with artificial light which has been discussed in earlier chapters helps give a building its particular character and quality. It is these qualities that set lighting for seeing and the perception of space apart from the purely servicing aspects of a building.

Level 2: integration of lighting with building services

Whilst the visual success of a building (level 1) must be the architect's first priority in satisfying the psychological needs of those using the building it would be idle to suggest that a person's physical needs can be ignored or left to other consultants. Whilst such matters may not be the immediate responsibility of the architect, the architect does have an overall duty to ensure that a person's physical needs are satisfied. There are many decisions to be made about services, where the successful integration of several factors will depend upon the architect.

The following is a list of building services which may impact on the lighting design:

Heating and ventilation/air movement
> Window design
> Heat dissipation/cooling
> Air conditioning
> Energy management/saving

Acoustic control
>   Acoustic isolation/external noise
>   Noise control within a space
>   Noise control between different rooms

Fire control
>   Sprinklers
>   Light fittings and fire control

Partitioning
>   Flexibility and light fittings

Loudspeakers
>   Public address systems
>   Telephones

Electrical distribution
>   Track
>   Trunking

Dimensional co-ordination
>   Integration

Each of these aspects of building servicing will be discussed briefly, not to suggest solutions, but to illustrate the way in which decisions on lighting can be affected by decisions on building servicing.

HEATING AND VENTILATION / AIR MOVEMENT

Daylight and artificial lighting strategies are both related to the needs of heating and ventilation and the architect needs to be aware of the consequences of the decisions he takes.

First to consider is daylight and with it solar gain. The entry of sunlight within a building, despite its positive values, brings heat and can cause thermal discomfort at certain times of year and possibly the need for additional cooling loads.

The architect's first line of defence is the orientation of the building to ensure that the window area needed to provide a high level of daylight does not at the same time increase the heat from the sun. This can be achieved by an east/west orientation. Alternatively heat protection can be provided by various forms of external shading, the second line of defence. Summer sun is at a higher level than in winter when the external shading may be designed to permit solar gain to reduce the heating load in the building.

Secondly there is the question of adequate ventilation, which works together with daylight to assist in providing acceptable environmental conditions. Ventilation raises the questions of occupant control and whether forms of mechanical ventilation are desirable. This brings up the issue of air conditioning and whether the ventilation is to be controlled by the Building Energy Management System (BEMS).

The entry of daylight is closely associated also with the nature of the glazing, whether single, double or triple, or one of the high-tech glass solutions. It is important here that the controlling glass chosen does not distort the colour or reduce the impact of the view to the world outside.

Having considered the role of natural light, it is necessary for the architect to take into account the impact of the artificial light on the needs of heating and ventilation. This is primarily a question of the heat generated by the lighting equipment, or its use of energy. The use of lighting energy at night is to some extent fixed by the function of the space, in that the type of light source best fitted to fulfil the function of the building, whether hotel or factory, will determine its total energy use.

The use of lighting energy during the day however, is very much within the control of the architect; since in the majority of architectural programmes, this will be determined by the daylighting design. In a low-energy environment much or all of the functional light can be provided during the day by the natural source, the energy otherwise used by artificial lighting equipment being significantly reduced.

The ceiling is a logical place for the location of lighting equipment, whether recessed or suspended, and the ceiling also provides a location for the heating and ventilation; indeed all three can sometimes be combined, with air being drawn through light fittings to remove excessive heat or air being forced through the apertures formed by the lights into the space beneath.

It is not the purpose here to suggest solutions to the problems of the heating and ventilation of a building, but to point out the paramount necessity for the architect to keep a grip on the solutions offered to ensure an elegant and integrated system.

ACOUSTIC CONTROL

There are a number of acoustic problems in a building which need to be addressed by the architect and his acoustic advisor. All have implications with other disciplines and merit an integrated approach, in which the lighting design may become involved.

The more obvious problems are as follows:

1 Internally-generated noise and the passage of unwelcome sound within a large space. The passage of sound from one space to another. The attenuation of sound by surface materials.
2 Externally-generated sound, from road traffic or aircraft, for example, associated with the building envelope and consequently with the introduction of natural light and air.
3 Noise generated by the lighting equipment itself; this may derive from the necessary control gear for discharge sources, or from the controls associated with the Building Energy Management Systems (BEMS).
4 Special acoustic design will be needed for those buildings, such as theatres and concert halls, where acoustics and lighting have an important and often interdependent role, and where the expertise of an acoustic consultant will be vital to the success of the building function.

Dealing with each of these issues in turn:

1 Internally-generated noise
The ceiling has been the traditional location for acoustic material for dealing with the noise from machines and people in a working environment, and this worked well until the size of lighting equipment placed on the ceiling reduced the area available, to the point when luminous ceilings took over, leaving little of the remaining ceiling for acoustic control.

Even though luminous ceilings may have lost their appeal, the need for acoustic control is still there, particularly with exposed concrete soffits, which may be exploited for environmental reasons other than pure structure. It may no longer be appropriate to add suspended ceilings (the catch-all answer to acoustic attenuation) and in situations where the runs of lighting equipment are suspended below flat soffits the solution may lie in associating the acoustic material with the design of the lighting equipment as shown in Case Study 13 on the Powergen offices in Coventry.

The absence of air conditioning in our new non-air-conditioned buildings can present a problem not of noise, but the lack of it. This is not a problem which can be solved by the lighting design, but it may well be a problem for the architect, and advice needs to be sought as to the introduction of white sound.

A further problem of internally-generated noise relates to the introduction of partitioning and the passage of sound through lighting equipment from one space to another. This is dealt with later.

2 Externally-generated sound
This is a problem easily solved in windowless buildings, but this is no reason for such buildings except in exceptional circumstances, such as in a concert hall or theatre.

The passage of sound through the window is associated with the passage of air and the means of ventilation. Buildings close to airports or on major traffic arteries will not only present noise problems but also problems of air pollution which must be addressed at the window.

Normal window glass transmits noise, but having established the window area required for all the less tangible but vitally important reasons fulfilled by the window, much can be done to ameliorate the problem of noise through it. This is very much the concern of the architect, and solutions to be investigated will no doubt include, among others, double or triple glazing and wide gap solutions with acoustic material around the edges. The architect's concern must however be with the prime function of the window to admit daylight and view.

3 Noise generated by the lighting equipment itself
This is a matter of specification, and it is likely that the most inexpensive solution will be the noisiest solution. In the past unacceptable noise level was produced by the controls needed to manage a lighting system. High-quality equipment is now far less noisy and provided that the specification contains limits for the sound generated at critical frequencies the architect should be able to have confidence that the lighting installation and its control systems will be acceptably quiet.

4 Special acoustic design
Whilst this is a specialist subject and not one for detailed discussion in a work of this nature it is important that the acoustic consultant liaises with the lighting designer since it may well be that there is a conflict between the shapes, materials and location of the acoustic control and that of the lighting design – the architect must be aware of any such conflict to ensure that in the final solution both conform to his overall design concept.

FIRE CONTROL

In the United Kingdom all buildings, including domestic dwellings, are controlled by legislation, which requires them to satisfy the building regulations as interpreted by the Local Authority fire officer. The basic requirement of such legislation is 'to offer the occupants of a building time for safe evacuation, free from fire and smoke.' On satisfaction of these requirements a Fire Certificate will be issued. For certain designated buildings open to the public, such as hotels, the most stringent regulations apply.

The relationship of fire control to daylighting is in the design of the building envelope and detailing of the apertures, but where artificial lighting is involved there are basically three forms of fire control which must be taken into account by the lighting design:

1 Fire resistant barriers;
2 The sprinkler system, required in certain categories of building;
3 Smoke detectors.

Where a building element such as a ceiling or wall is designed to act as a fire separation or fire barrier any lighting equipment recessed into the fire barrier must have the same fire protection as the building element itself. In practice this means that where lighting equipment is recessed into suspended ceilings acting as fire barriers, the specification for the light must include the same fire protection as that provided by the ceiling.

Solutions have been found in the case of certain fire resistant ceilings such as fibrous plaster, where the light fitting has been designed from the same fibrous material as the ceiling, but this is not a solution which finds favour with lighting equipment manufacturers

In the case of a sprinkler system designed to suppress fire, when flames have reached between a third to half of the ceiling height water is deluged into the space from a regular array of apertures or sprinkler heads in the ceiling fed by an overhead pipe system. The layout of these apertures will take priority over the ceiling layout so that the location of lighting equipment recessed or close-mounted to the ceiling needs to be co-ordinated with the sprinkler system, if an elegant solution is to be adopted. This should not prove a difficulty and generally systems of artificial lighting and of fire control can be co-ordinated satisfactorily unless left until too late.

Smoke detectors rely upon smoke particles reaching them. If lighting fittings produce significant amounts of heat which collect below a flat ceiling soffit this layer of heat may prevent the smoke from reaching the detector and consequently stop the early detection of a fire. The lighting designer should be aware of this possibility, which may be solved by the re-siting of the detectors.

When the lighting of an existing building is modified it will be increasingly necessary to ensure that fire safety is not endangered. A further aspect of fire safety which must be borne in mind is the means of escape and this is a science in itself in which clearly lighting has a large part to play.

PARTITIONING

The relationship between lighting, acoustics and partitioning has already been discussed, and there is little that needs saying about the relationship of lighting with partitioning where questions of acoustics are not an issue.

However early co-ordination between the lighting designer and the architect is needed to ensure that the notional lines of partitions designed to reach the ceiling, whether structural or suspended, are made known to the lighting designer, so that when the layout for the lighting design is planned, there is no conflict with the position lines for the partitions which are to installed initially, or may be planned for the future.

The present tendency is for flexible office layouts with low level partitions, and here the question is more the layout of floor sockets to service the many needs of computers and communications systems which need to relate to changes in the office layout. Where low level lighting at the control of the occupant is desired, the nature of the lighting equipment needed may well be a part of the partitioning package, although its control needs to relate to the overall control system of the space.

LOUDSPEAKERS

The relationship between loudspeakers, public address systems, and other means of communication have only a tenuous relationship with the lighting design, although where telephone control of light fittings is planned their relationship needs to be considered.

For the most part, the problem is one of co-ordination of elements to ensure elegant solutions to the visual experience, where all items in the ceiling are planned to live happily together, and where the servicing of one element does not prejudice another.

ELECTRICAL DISTRIBUTION

Lighting design is of course inseparable from electrical distribution. The general electrical distribution for the building will be the responsibility of the electrical consultant who will be involved in the supply of power to all its many needs.

Electrical power for lighting is an important need and one that constitutes a major energy use. It is important, therefore, that the parameters for the lighting design are set out at as early a stage as possible to assist the electrical engineer in his planning work. It will be the fault of the architect if the ideal lighting concepts planned by the lighting designer are left so late that the electrical layout has made no provision for the factors involved. Conditions may be such that it would have been impossible for the engineer to leave his design decisions to a later stage.

DIMENSIONAL CO-ORDINATION

It was said in 1964 that 'Development in the field of building servicing should lead to a much closer integration of servicing requirements, to rationalise the multiplicity of present solutions into a single entity designed to satisfy all our environmental needs. The relationship of building structure to building servicing should be reconsidered in order to discover if the structure can be designed to contribute an increased performance in servicing, and to encompass some of the services at present satisfied in other ways.'[1]

[1] Phillips, Derek, *Lighting in Architectural Design*, McGraw Hill, 1964.

There are now many examples of where this has been done, with structural slabs used for cooling and the building envelope designed to cope not only with the needs of natural lighting and sun control but also with acoustics, ventilation and heating.

There is a relationship between savings in energy and integration of services and the future will no doubt see further advances in this field, advances which it is to be hoped will be architect-led.

Part 2

Introduction to Case Studies

Part 2 of the book consists of 59 Case Studies of Buildings outlining the development of both daylighting and artificial lighting throughout the twentieth century from the early 1930s to the Millennium.

The buildings chosen vary from those early factories where the needs for artificial lighting combine with those of the natural source to provide an integrated solution. Buildings are as diverse as the Bexhill Pavilion built in the 1930s, the Brynmaur rubber factory in Wales built at the end of the war, or the recently completed Hong Kong International Airport.

The projects chosen are representative of the solutions which have been adopted, and where coherent strategy is evident; in which the needs of natural lighting have been combined with those of the artificial source.

This is not to deny those solutions in which the needs for a controlled artificial environment for certain architectural programmes, have appeared to merit the total elimination of daylight; but there is a clear recognition towards the latter part of the Century of the very real advantages of natural light, associated with a world shortage of energy, leading to solutions in which the qualities of both natural and artificial light are designed to complement each other.

Within the 59 projects a wide variety of architectural programmes has been included, from Homes and Churches to Hospitals and Airports; whilst an attempt has been made to include some of the more difficult problems in the form of the display of works in Art Galleries.

It is not intended that the inclusion of any project is, in itself, some form of 18th Century 'pattern book solution,' but rather that there are lessons to be learned from the strategies which have been adopted.

It is for this reason that technological details of the artificial solutions have been kept to a minimum, these are in any case in a constant state of evolution; it is in the 'quality' of the strategy that its importance lies.

Section 1 Residential

Case study 1 House in Chelsea, London

Sections and plans of the house (Derek Phillips Associates).

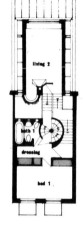

LEVEL 1 **LEVEL 2** **LEVEL 3** **LEVEL 4** **LEVEL 5**

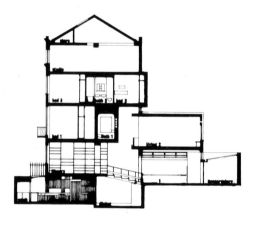

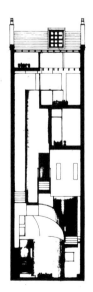
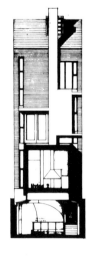

This conversion within a row of Regency terrace houses built in the early 1970s might at first sight seem to fit more aptly into *Lighting Historic Buildings* (Phillips, Derek, McGraw Hill, 1997) but apart from the original front wall to the street (required to be preserved to maintain the integrity of the street frontage) the whole of the remainder of the house has been rebuilt, so it is therefore considered a new house.

The architects, Derek Phillips Associates, were briefed by the client Bernard Stern, director of a light fittings manufacturer, to 'fill the house with daylight' during the day and to exploit all the latest lighting technology for the artificial lighting at night.

A study of the section indicates the way in which light from the street side enters through the tall windows of the period, while at the rear the open plan allows natural light to permeate into every space in the house, except the small central bathrooms.

A decision was made to build over the original garden enclosing the entire site, ending with a small top-lit conservatory. Two living rooms, one poised on top of the other, allow daylight to infiltrate between the two; and even the basement kitchen derives daylight from overhead glazing.

Two staircases provide the vertical circulation, the one encircled by the other. The inner staircase allows staff and children to move about the house without disturbing the main reception rooms and the basement dining room, and is crossed by a bridge to reach the ground level living room.

The house was built at a time when lighting track had just been introduced, and it has been used extensively to add flexibility to the lighting design, enabling the light fittings to be changed with changes of need, at a later date. This can be seen in the lower living room which has lines of track installed down each side of the space, between the ceiling and the overhead glazing. This has already proved valuable in permitting smaller, low voltage fittings to be applied as they became available.

The overall impression of the interior is one of light, both natural and artificial, which complements the owner's collection of works of art.

Architect and Lighting Consultant: Derek Phillips Associates

Lower living room during the day (Photographer Derek Phillips).

Exterior of Regency Terrace (Photographer Derek Phillips).

Lower living room at night (Photographer Derek Phillips).

Case study 2 Koshino House, Ashiya City, Japan

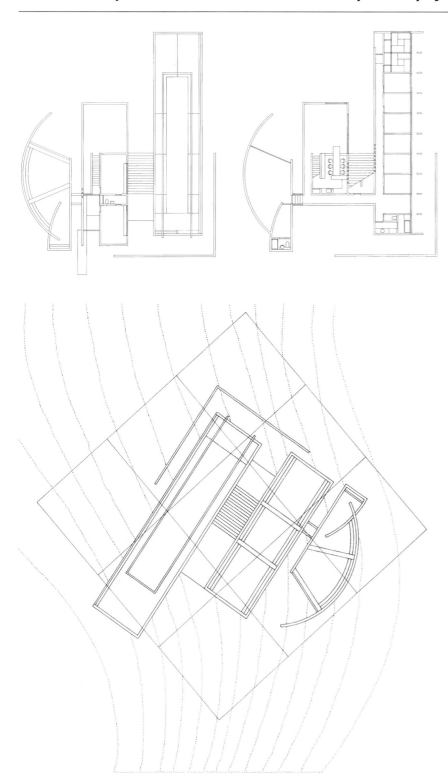

The Koshino House, designed by architect Tadao Ando, was completed in 1981, with an atelier added three years later. The house consists of two rectangular concrete boxes joined to each other to form a small courtyard. The smaller of the two boxes contains a double-height living room, kitchen and dining area, with master bedroom suite above, whilst the lower box contains six smaller rooms for children.

The house is located at the foot of the Rokko mountains, in Ashiya City, east of Kobe with wooded countryside falling away in a gentle slope, allowing views out from the main living area. The house is approached from above, where a visitor views the roofs which express the plan form of the house and the circular atelier.

The interaction of daylight with fairfaced concrete walls plays an important part in all of the architect's work, which has given him the justly-earned accolade of 'architect of light.' The Koshino House is no exception to his work. As can be seen in the pictures, the whole form of the building interior expresses its relationship with daylight and view, sunlight and shadow.

Plans of the house (Tadao Ando).

Architect: Tadao Ando

Aerial view of the completed property (Photographer Hiroshi Kobayashi).

Double-height space overlooking the sloping ground between the two blocks (Photographer Tadao Ando).

Night exterior of the living room and its relationship with the two blocks (Photographer Mitsuo Matsuoka).

Case study 3 House in Islington, London

Ground and first floor plans to show the use of the spaces (G. Jury).

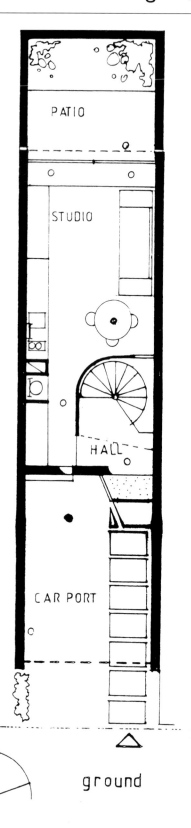

PATIO

STUDIO

HALL

CAR PORT

ground

Built on three floors with a basement, the house built by designer Gerald Jury has a narrow frontage of only 3.6 metres and is fitted into a row of mews houses in Islington in London. Every part of the 17-metre-long site has been developed.

The front of the ground floor is a car port and the rest is a studio flat for the designer's daughter, a sculptor; and as can be seen in the section this looks on to a small patio which provides an extended view, not unlike Japanese houses where space is at a premium. The basement provides space for a sauna, utility area and sculptor's workshop.

The two upper floors are where the designer and his wife live, with their own outside terrace at roof level. The planning makes the most of the space available and despite the narrow site natural light and view are brought in to all the living areas. A glass conservatory at roof level allows natural light to enter the vertical circulation, a circular staircase, so that even in the centre of this deep plan, the house appears to be full of daylight.

The house is a fine example of the designer's innate sense of the need for natural light in the home, where no amount of daylight calculation would have been necessary.

Little needs to be said about the artificial lighting, which is subtle and admirably fits the rooms, using mainly portable types of fitting. Some recessed filament fittings are added, where more functional light is required.

Designer: G. Jury

Long section (G. Jury).

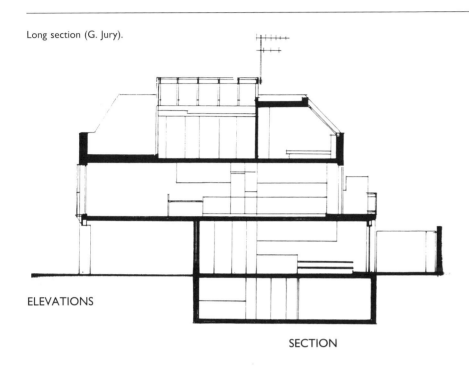

ELEVATIONS

SECTION

First-level dining kitchen with its view to the terrace (Photographer G. Jury).

Second-level landing, with daylight shadows (Photographer G. Jury).

Conservatory rooflight at high level with view to terrace (Photographer G. Jury).

Case study 4 High Cross House, Dartington

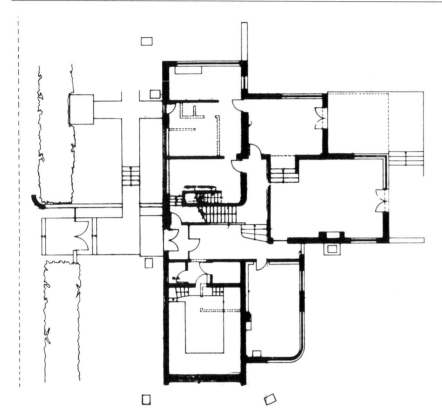

Floor plans of the conversion (John Winter).

Exterior photograph (Photographer Derek Phillips).

Dorothy and Leonard Elmhurst bought the Dartington Estate in 1925 to start the Dartington Hall experiment in education and the arts. In 1930 the headmaster W. B. Curry brought in the Swiss-American architect William Lescaze from Philadelphia, who built a series of houses and other buildings in line with the International style of the Modern Movement in 1932, which included High Cross House for the headmaster himself.

The house, one of the finest examples of the International style in Britain, had all the ideal characteristics of daylighting and colour with splendid views out to the countryside beyond, which are summed up by W. B. Curry himself who wrote at the time: 'To me, serenity, clarity and a kind of openness are its distinguishing features, and I am disposed to believe that they have important psychological effects upon the occupants.'

In 1991 a decision was made to convert the house into a gallery for the display of the Dartington collection of paintings and ceramics, and this presented the architect, John Winter, with a conservation problem, since a house so full of daylight did not provide the most conducive atmosphere for the display of paintings. This has been solved by using the reception rooms on the south side for their original purpose and converting the service rooms with their small windows on the north side for the display of paintings. The small windows were provided with ultraviolet filters and venetian blinds which are closed when the gallery is not open to reduce the number of hours of daylight on the paintings.

The pottery gallery entered off the main entrance takes advantage of the daylight from the large south-facing windows, since there is less of a conservation problem here. Lighting track is placed on the ceiling to provide downlighting to the glass cabinets, which are also lit by artificial light from below.

The most dramatic change in the building is the conversion of the original garage into a double-height gallery for temporary exhibitions, with a mezzanine level. Daylight is excluded, and the flexibility needed for an exhibition space of this type is provided by tungsten halogen spotlighting from ceiling-mounted track.

Architect: William Lescaze; Refurbishment Architect: John Winter

Pottery room (Photographer Derek Phillips).

The artificial lighting of the original building has been replaced as far as possible to echo the original effect using a match for the original glass hemispheres, but with long life lamps. Some simple uplighters are used with angle poise table lamps.

Temporary exhibition (Photographer Derek Phillips).

Section 2 Ecclesiastical

Case study 5 Clifton Roman Catholic Cathedral, Bristol

The exterior view, showing the soaring roof structure (Photographer Derek Phillips).

The stained glass wall (Photographer Derek Phillips).

Built in the late 1960s, the Catholic Cathedral in Bristol, designed by architects Sir Percy Thomas and Partners, was a landmark building in which a unity exists between the design for the reinforced concrete structure and the introduction of natural light.

The tall, central roof structure has been designed to allow daylight to flood into the building, emphasizing the chancel area. At the same time the lower, coffered roof structure stops short of the walls to allow natural light to permeate through the gap; a device which has also been used to conceal the artificial fluorescent sources from view, which illuminate the walls with their sculptured stations of the cross.

The whole interior glows with light, both during the day and at night, with the light always coming from the same direction so that no significant changes occur as the daylight fades. However, when a church service is planned after dark, the emphasis of the artificial lighting is directed onto the chancel area, creating a change of mood. During the day variety is provided by the changing exterior conditions, whilst at night it is provided by changes in the artificial lighting.

A wall mural of stained glass provides a dramatic edge to the area containing the font which itself is lit by a smaller overhead daylight roof opening.

The use of natural and artificial lighting in Clifton Cathedral were both state of the art in the 1960s, and the building was noted both for its low cost and its integrated interior.

Architect: Sir Percy Thomas Partnership; Services and Lighting Design: Roger Preston & Partners

The font area with overhead roof light and stained glass wall (Photographer Derek Phillips).

Ceiling detail allowing the entry of both daylight and artificial light (Photographer Derek Phillips).

Case study 6 Church of the Light, Osaka

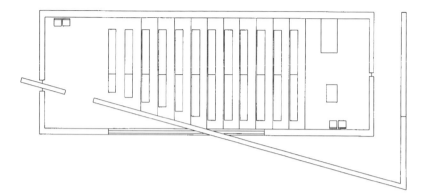

Plan of the church (Tadao Ando).

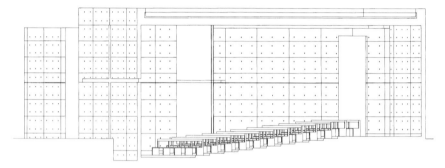

Long section (Tadao Ando).

Exterior (Photographer Mitsuo Matsuoka).

The Church of the Light, designed by architect Tadao Ando for the Ibaraki Kasugaoka Church, a member of the United Church of Christ in Japan, is yet another example of the work of a master of the manipulation of natural light. The building is quite small, only 113 square metres, built in a quiet residential suburb of Osaka.

The church is a rectangular, concrete box, intersected at 15 degrees by a free-standing wall (see plan). The most outstanding feature of the church interior is the ceiling-high cross cut out of the wall behind the altar. Light penetrates the darkness of the concrete box throwing a pattern of light on to the floor.

This opening was originally to be left clear to admit both daylight and air, but the practicality of climate determined that it should be filled with glass. This in no way diminishes the impact of the 'cross of light' which enters the interior.

The type of artificial lighting used can be seen in the photograph and is clearly a minimal solution, consisting of simple wall brackets on one side wall, directing light downwards.

The furniture is made from rough scaffolding planks, as is the floor, which slopes down towards the altar. The architect believes in the 'use of natural materials, for parts of a building that come into contact with people's hands or feet that have substance, since it is through our senses that we become aware of architecture.' If this is so then it is through our emotions that we are aware of light.

Architect: Tadao Ando

Interior towards the rear showing the pattern
of the cross (Photographer Mitsuo Matsuoka).

Interior towards the altar end (Photographer Mitsuo Matsuoka).

Case study 7 Extension to village church, Aldbury

Ground plan of
church and extension
(Peter Melvin).

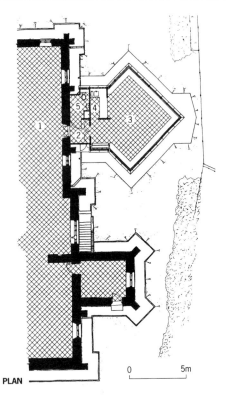

PLAN

0   5m

Axonometric to
show relationship
(Peter Melvin).

NORTH EAST ELEVATION

Whilst alterations to historic buildings generally are covered in the author's earlier book, *Lighting Historic Buildings*, Butterworth-Heinemann/McGraw Hill, 1997, this small extension to a fourteenth century church in the village of Aldbury in Hertfordshire is such a good example of how such things should be done, that it deserves a place in these case studies for new works.

The brief to the architect Peter Melvin, who lives in the village and understands its texture, was to provide extra space for small meetings, Sunday school, choir practice and social events.

Aldbury is a conservation village, where all new building is scrutinized carefully to ensure that it will fit. This church extension was no exception, and by placing the new plan on a diagonal to the main structural theme of the tower, and echoing the flint and stone construction, the new work sets up a dialogue with the old which gives the appearance of it having always been there.

The daylighting is from the wrap-around windows which offer splendid views to open countryside across the churchyard. They are made of clear double glazing, with side panels of blue glass to modify the interior. Whilst physical contact with the original church is kept to a minimum, the diagonal form allows visual contact with the main structure.

The modest scheme of artificial lighting is provided by a series of wide-angle spotlights mounted on the roof beam structure; no attempt has been made to integrate them with the pyramidal ceiling. It is a lighting design which admirably fits the subtle simplicity of the architecture and provides all the electric light required to satisfy the brief. Lighting control is by simple dimmers, giving all the flexibility required.

Architects: Peter Melvin & Jane Newman of Atelier MLM

Exterior (Photographer Derek Phillips).

Daytime interior (Photographer Dennis Gilbert).

Case study 8 Bagsvaerd Church, Denmark

Section

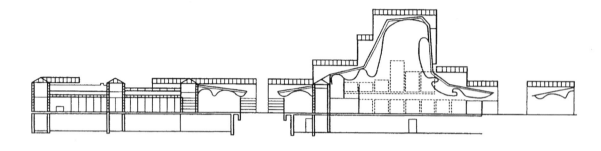

Plan (Architect Jorn Utzon).

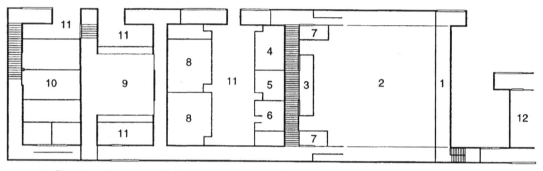

1. Church porch
2. Church-interior
3. Vestry
4. Choir and organist
5. Parish clerk
6. Vicar
7. Storage room
8. Confirmation-rooms
9. Meeting hall
10. Youth center
11. Back gardens
12. Chapel

Church exterior (Photographer Kit Cuttle).

The architect Le Corbusier was reputed to have said that a 'church is a gadget for holding service.' He also described a home as 'a machine for living in.' But he then proceeded to ignore both remarks in his work at Ronchamp and elsewhere.

Whilst the architect for the church at Bagsvaerd is said to have noted Le Corbusier's remark, happily he too ignored it. In doing so he has produced a building where thin concrete shells have been used to create the impression of a ceiling of clouds dissolving upwards to an invisible sky. It would be difficult to imagine anything less like a gadget; indeed the interior is closer to baroque in the way that large windows concealed from normal views allow natural light to flood into a space.

The church at Bagsvaerd in Denmark was completed in 1976, having had a long gestation period reaching back before the First World War. By any standards it is a large parish church, measuring 80 metres by 22 metres wide and designed to accommodate a congregation of 350 people.

Architect: Jorn Utzon, Denmark

Interior towards the organ (Photographer Kit Cuttle).

View across the church (Photographer David Loe).

The characteristic impression of the church comes from the concealed window 16 metres above the floor where two large shells overlap, the 'light changing during the day from light grey to golden, on the arrival of the afternoon sun pouring into the space – sometimes the drifting clouds are reflected in the large vault above the altar.' Elsewhere, skylights above the aisles and vestry allow the changing colours of light to fall into the church, at times all purple or blue, whilst at night the twinkling of the many glow-lamps gives a soft subdued lighting effect.

The interior of the concrete walls and the insides of the shells are lime-washed in white and the floor is composed of white concrete tiles. Colour is provided by the occupants and the vestments of the clergy.

From the exterior the church may not look like the everyday churches of the nineteenth century, but this is a church of our time, where the visual effect of the interior harks back to the great traditions of the past.

Case study 9 Fitzwilliam College Chapel, Cambridge

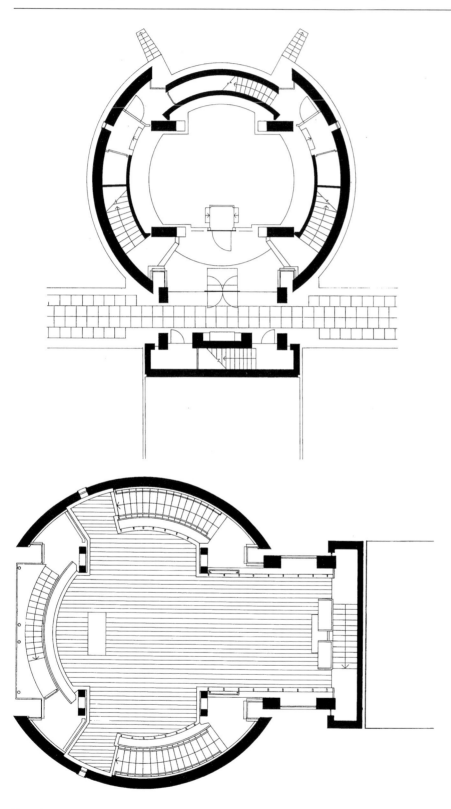

The Fitzwilliam Chapel of Cambridge, completed in 1992 and designed by Richard MacCormac, is an example of the richness of experience that can be obtained within a daylight strategy, not dissimilar to that of the baroque churches of southern Germany.

By separating the curved side walls from the roof, daylight is fed into the perimeter of the space, disassociating the central congregational space from the circulation.

The accommodation is on two levels: the crypt or meeting room at low level; and the chapel itself poised above like an ark, in the words of the architect 'symbolising the concept of passage and protection – the ship signifying the way of salvation.'

The congregation faces a large window to the east behind the altar, overlooking a large plane tree. The balance between the top light and east light in the chapel varies throughout the seasons, providing a continuous sequence of experience.

The artificial light derives from fittings recessed into the ceilings in an ordered pattern relating to the cruciform ceiling design. At night a very dramatic view of the chapel is gained from the garden outside.

Two plans

Architect: MacCormac Jamieson Prichard

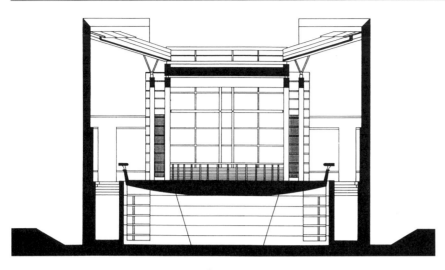

Cross-section (MacCormac Jamieson Prichard).

View up staircase to chapel (Photographer Michael Evans).

Exterior view at night (Photographer J. J. Photoservices).

View up staircase to altar end, showing east window (Photographer P. M. Blundell Jones).

Section 3 Offices

Case study 10 BA offices, Waterside, Heathrow

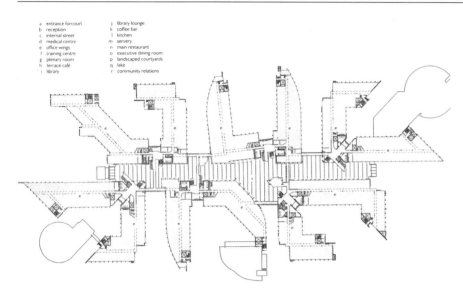

a entrance forcourt
b reception
c internal street
d medical centre
e office wings
f training centre
g plenary room
h terrace café
i library
j library lounge
k coffee bar
l kitchen
m servery
n main restaurant
o executive dining room
p landscaped courtyards
q lake
r community relations

typical upper level plan

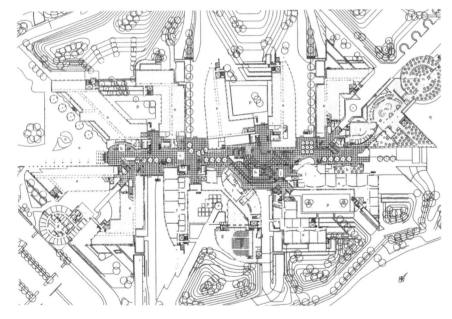

Floor plan of whole building complex (top only) (Niels Torp).

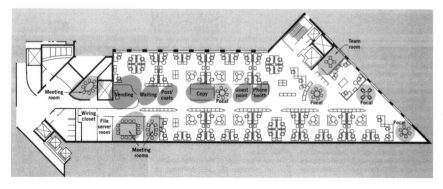

Floor plan of individual, open plan office (Niels Torp).

The British Airways offices at Waterside close to Heathrow Airport are as much an experiment in working as they are in environmental control. Set in 270 acres of newly-created park land at Harmondsworth, the two hundred million pound office complex consists of six separate buildings linked by a glazed street to give office accommodation to some 2800 staff.

The buildings, designed by Norwegian architect Niels Torp, are similar in concept to those he designed for the SAS airline in Sweden, and rely on the provision of generous working conditions in open plan offices, with all grades of staff having the same open desk layout. The street provides access to cafés, shops, a supermarket and other normal high street facilities, making it unnecessary for staff to leave the building during the day. The building is open twenty four hours a day.

The buildings are air conditioned, whilst the street is not, allowing the temperature to rise in the summer to reflect the exterior environment. The street has a glazed roof, sealed against the weather and outside noise. The offices are close to Heathrow Airport, although not in the flight path. They are provided with cross ventilation to ensure that the temperatures are not excessive. The glass is treated to cut down on solar gain.

The offices in the six separate buildings are essentially daylit, with supplementary artificial light being obtained from recessed fluorescent luminaires. The impression on visiting the offices is of natural light dominating the street from the overhead glazed roof. Daylight enters the four floors of offices both from the street and from the glazing along the exterior walls.

The fluorescent light fittings are controlled by a system of presence detectors which react to the daylight available outside. At all times of the day when a person arrives on an unoccupied floor the presence detector will automatically switch on the lighting of the particular zone.

The lighting control system for the offices is designed to ensure that whatever the level of daylight outside, the offices are not overlit, the artificial lighting level being adjusted to provide 450 lux and no more, from the combination

Architect: Niels Torp, Norway;
M&E Consultant and Lighting Design: Cundall Johnson & Partners

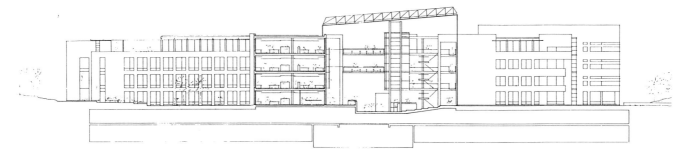

Section through complex to show the relationship of the street to the separate buildings, showing the roof glazing (Niels Torp).

Exterior view showing the entrance (Photographer Derek Phillips).

of natural and artificial light. The target light level has been set at 450 lux in the offices, but where the use of a space is changed this level can be reduced. Additional suspended fittings are placed in circulation areas giving an element of upward light.

Whilst the office planning provides for teams working in open areas segregated by moveable screens, each floor is provided with a series of smaller enclosed working areas and conference rooms that use all the most modern methods of information technology.

View of the street in the day (Photographer G. Price, BA copyright).

View of the street at night (Photographer Thorn Lighting).

Case study 11 Solar office, Doxford International Business Park

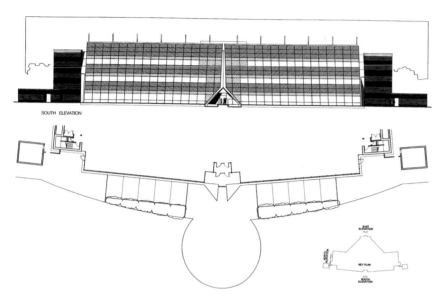

Elevation of building (photography Studio E).

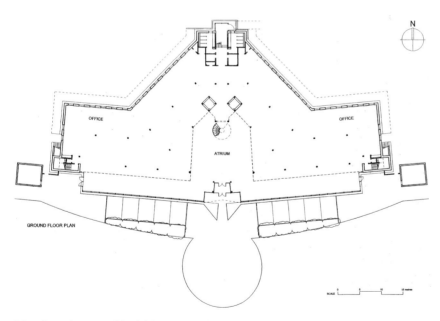

Office floor plan, ground level (photography Studio E).

The Solar office at Doxford International Business Park is one of the first serious attempts in the United Kingdom to use energy from the sun to assist in providing electrical power for a building. Studio E Architects have designed a south-facing façade incorporating a photovoltaic (PV) array to catch the sun and convert it to useful electricity.

In the past the use of PVs has been of a specialist nature and has not been considered commercially economic. It is therefore all the more surprising that the client for the building, Akeler Developments plc, a company primarily concerned with speculative office development, has considered that taking the whole life of the building into account, the energy saved by using PVs has commercial viability.

The brief to the designers, worked out by Studio E was 'To provide a distinctive building, to contain versatile and responsive office space, to ensure that the building makes minimal impact on the global and local environment and to integrate within its fabric the means to generate a worthwhile proportion of its energy needs.'

The 15-metre-tall PV façade supplied by Schuco International slopes back at an angle of sixty degrees to maximize the solar radiation exposure whilst ensuring that glare reflected from the wall is not detrimental to the vision of drivers on a neighbouring trunk road.

The energy used for lighting in an office building is a significant proportion of the total energy use, and the PV wall is designed to replace a significant amount of that energy by providing 55 000 kilowatt hours per annum (an energy generation target of 73 kWp).

The quantity and quality of daylight was central to the Doxford project. The offices have been designed to eliminate the use of artificial light on normal days, and it can be seen from the section that daylight enters from the south side between the PV bands and from the north by means of vertical glazing. This provides an average daylight factor of 2 per cent over 80 per cent of the office space. In addition, the central atrium has

Architect: Studio E Architects; Co-ordinating Architect: Aukett Associates; Building Services: Rybka Battle

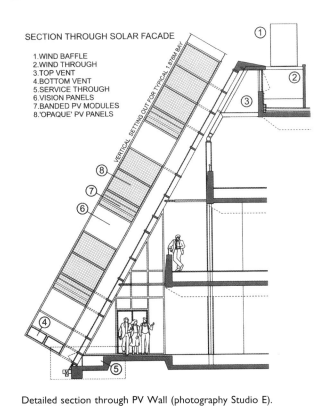

SECTION THROUGH SOLAR FACADE

1. WIND BAFFLE
2. WIND THROUGH
3. TOP VENT
4. BOTTOM VENT
5. SERVICE THROUGH
6. VISION PANELS
7. BANDED PV MODULES
8. 'OPAQUE' PV PANELS

VERTICAL SETTING OUT FOR TYPICAL 1.876M BAY

Detailed section through PV Wall (photography Studio E).

roof lights which permit natural light to enter and which also double as smoke vents.

The building was completed in 1998, and as the first speculative solar office, it is being closely monitored to check whether the PV façade will produce the anticipated 25–33 per cent of the office's electricity needs, and that the office space conforms to the desired comfort targets.

The building is as yet unoccupied and the artificial lighting is to be installed as part of the fit-out to satisfy the occupiers' requirements. However, it is likely to be a system of low-brightness fluorescent fittings giving 300–400 lux with either simple manual or automated controls.

Exterior to show PV wall (photography Studio E).

Interior looking towards the PV wall, showing the daylight effect (photography Studio E).

Case study 12 BRE environmental building, Garston

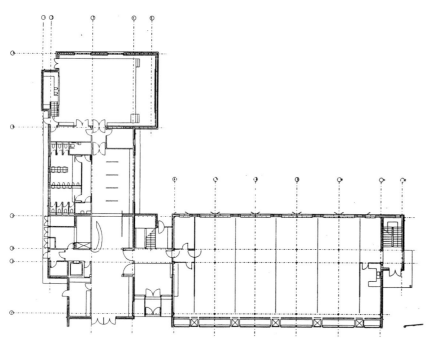

(a) Plan of the building (Feilden Clegg Architects)

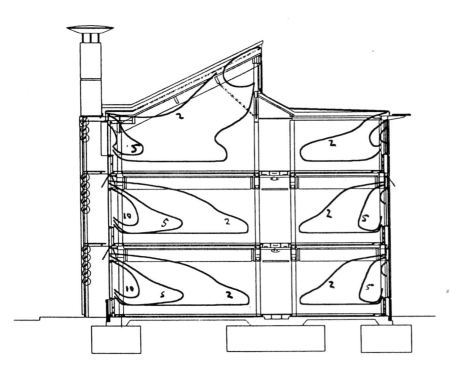

(b) Cross-section through the building to show the daylight factors (Max Fordham & Partners)

In a similar manner to the low energy office at Doxford, the low energy office at the Building Research Establishment in Garston set out to solve the lighting problems in an office in a low energy environment whilst also bearing in mind the need to solve the associated problems of ventilation and air movement without using air conditioning.

In the case of the BRE, it was intended that the building would be both a showcase and a test bed for further research, diminishing the importance of the economic imperative normally associated with a speculative office.

The brief to the architects, Feilden Clegg and Max Fordham, the M&E services engineers, was to provide 'a comfortable and healthy internal environment' applicable to greenfield sites. The brief covered areas such as energy consumption, comfort and visual environment. In addition, the office space should be flexible to allow for the future movement of partitioning.

The general accommodation to be provided was as follows:

1300 square metres of office accommodation for 100 staff; 800 square metres of seminar rooms, with one main room and two smaller.

The main accommodation is arranged on three floors with a main office (approximately 30 metres by 13.5 metres) at each level. (see (a) floor plan)

The aim was to provide a minimum of two per cent daylight factor over the office area, and this can be seen in the section, achieved by fifty per cent of window area on the two side walls (see (b) cross-section). Solar shading is provided on the south exposure using glass louvres designed and controlled electronically to cut out sun glare but allowing maximum daylight to the office space, to reduce where possible the need for electric light during the day.

The choice of glass was made after full-sized tests were carried out, using 10 mm toughened clear float glass with a white ceramic coating on the underside of the

Architect: Feilden Clegg Architects;
Services and Lighting: Max Fordham & Partners

(c) Exterior of the south elevation showing the louvres (Photographer David Loe).

(d) Top floor. Unoccupied without artificial lighting (Photographer Derek Phillips).

(e) Detail of wavy ceiling and lighting, related to view (Photographer David Loe).

louvre; this excludes sufficient solar heat gain, whilst allowing some daylight through. The light transmission of the louvres is forty per cent with a reflectance of fifty per cent.

The solar louvres are controlled electronically by BEMS and are related automatically to the level of the daylight available inside the building.

The artificial lighting, which reacts to the daylight level and reduces to nothing when the daylight is sufficient, is designed to provide an illuminance level of 300 lux. It consists of slim lines of fluorescent fittings containing high frequency lamps. The distribution from these fittings provides for forty per cent upward light and sixty per cent downward light.

Provision has also been made for task lights when higher levels of light of up to 600 lux are thought necessary, but with the use by most staff of computers, the need for higher levels of light is unlikely.

The lines of fluorescent fittings are set at right angles to the side walls, being suspended from the centres of the curved folded plate ceiling (see (e)) at ground and first levels. At the second floor level with its taller ceiling and clerestory daylighting the same type of fittings are suspended along grid lines.

It is too early to assess the overall savings of power in this low energy building but the target figure for the electrical load for lighting of 10 kW h/m^2 year has been achieved by the daylight strategy adopted.

Case study 13 Powergen, Coventry

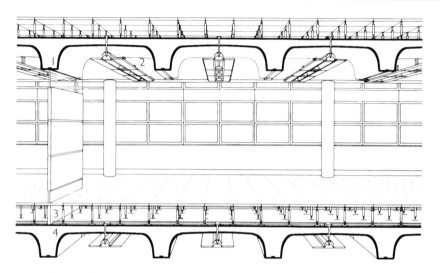

Section through the concrete soffits to show the relationship with the lighting fittings (Bennetts Associates).

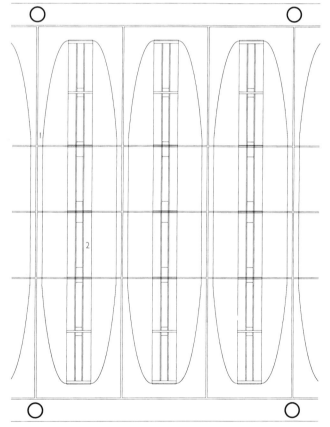

Typical bay plan

1 Rebate in Concrete for glass partition heads

2 Light-raft suspended from slab

3 Raised floor void accommodating all service runs

4 Coffered concrete structure cast- in situ

Plan of the lighting system (Bennetts Associates).

When Powergen was privatized in 1990, it inherited two offices in the West Midlands but decided that to increase efficiency it would house its 600 staff under one roof at the Westwood Business Park near Coventry. The architects, Bennetts Associates, were commissioned with the following brief in terms of the environmental conditions required:

to provide a modern efficient environment for safe and healthy working conditions;
the building to be energy efficient, with low running costs; and to encourage individual control of their own environment.

The building orientation is east/west, with natural lighting as a main requirement, no air conditioning and overnight cooling, and electrically-operated windows at roof level. Windows are double glazed, with sun shading on the south elevation.

One of the main features of the structure is that there are no suspended ceilings, the structure containing all the necessary services within it. This is expressed as an important design feature with the artificial lighting suspended in lines parallel to the bare coffered concrete soffits, giving light both up and down.

The fittings are specially designed using the compact triphosphor fluorescent lamp with high frequency control gear. The light distribution satisfies the requirements for VDU working, and the fittings are linked to the energy control system, reacting to the level of daylight available outside, so reducing the electrical energy otherwise used by the lighting system during the day.

In addition, the light fittings are adaptable for changes which might be required at some future date to the partitioning system. Another feature of the lighting system is that it had to incorporate acoustic attenuation. Control can be by telephone linking for individual control where desired.

Natural light is available from the window walls and from a skylight running the full length of the building, giving daylight to a central atrium. The tree-lined atrium is an important element in

Architect: Bennetts Associates; Lighting Design: Thorn Lighting

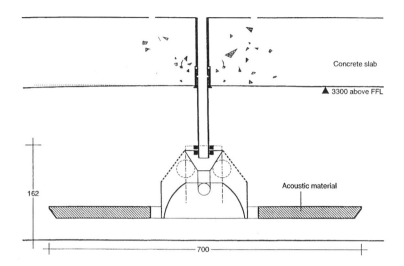

Concrete slab

▲ 3300 above FFL

162

Acoustic material

700

Detail of light fitting to show acoustic control (Bennetts Associates).

the plan, adding spaciousness, providing light from both sides of the offices.

The visual impression on entering the building during daylight hours is of a completely daylit building, despite the fact that when it becomes overcast the artificial light needs to be brought into play.

Cross-view to show light fittings (Photography Thorn Lighting).

Exterior view of south elevation (Photographer Derek Phillips).

The atrium (Photographer Derek Phillips).

Case study 14 Century Tower, Tokyo

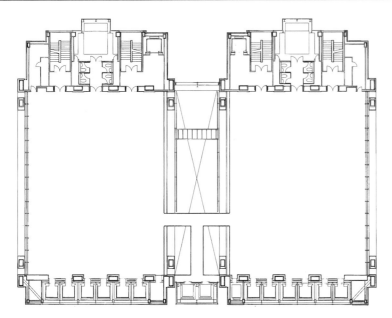

Plan (Foster and Partners).

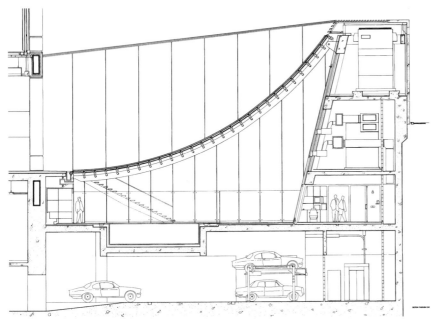

Section through building (Foster and Partners).

The Century Tower is an example of a large office block built as twin towers in central Tokyo. It was designed by Sir Norman Foster and follows the same design philosophy he applied to the Hong Kong and Shanghai Bank. Whereas the Hong Kong Bank was designed and built for a single client, the Tokyo Tower was designed as a speculative office block in which the various spaces could be sub-let.

This did not prevent the architect from adopting the same concept of a central atrium, through which natural light filters to the lower floors of the two towers, of nineteen and twenty storeys.

A key factor in the design, and of which there is little evidence from the outside, is the double-height floor system, with one suspended from the other. The structure expresses the two floors as one, which appears to reduce the mass of the building when seen at a distance. By locating the lifts and services in the two short sides of the office floors, natural light and view from the offices is unrestricted, which is evident from the plans.

Artificial light is of traditional form, using recessed fluorescent fittings placed in lines at right angles to the exterior windows and controlled by computer. However, the general impression on visiting the offices during the day is one of daylight, despite the supplementary artificial light which may be required.

The building provides for a wide range of activities and includes a restaurant, fitness centre and swimming pool; together with a museum and a traditional Japanese tea house.

Architect: Foster and Partners; Lighting Design: Claude Engle in association with Roger Preston & Partners

Suspended floors (Photographer Martin Charles).

Restaurant at night (Photography Foster and Partners).

Exterior of the tower, day and night (Photographer Ian Lambot).

Case study 15 Number One Regent's Place, London

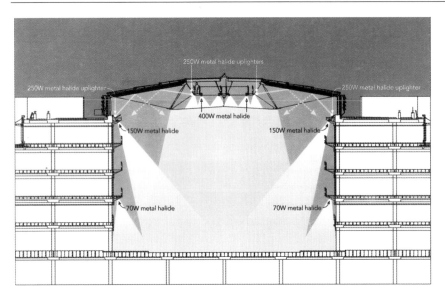

Schematic to show daylight and artificial light (Arup Associates).

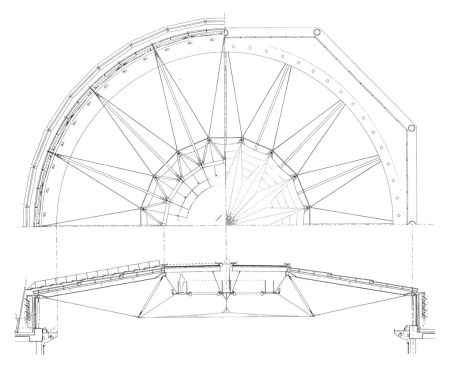

Detail of section through roof (Arup Associates).

Number One, Regent's Place, London is a 26 000 square metre office development designed by Arup Associates, consisting of five storeys of 18-metres-deep accommodation arranged around a central atrium.

A decision was made to develop the atrium as a trading floor for 300 dealers, and a specification was developed for the daylighting which may be summarized as follows:

Daylight control is to be afforded to the atrium roof glazing to provide 700–800 lux at the trading floor. The solution adopted is to meet the requirements for VDU operation. The daylight is to be managed by an automatic control system to maintain these illumination levels, with a view to optimizing energy consumption. Artificial lighting is to be provided within the atrium to supplement daylight to the same lux levels, when daylight is inadequate.

An important additional consideration when meeting these requirements was that the quality of the daylight as experienced in the atrium should remain uncompromized.

The daylight at roof level enters through a central occulus, and through clerestory windows around the perimeter of the atrium. The level of daylight penetration is controlled to ensure that it does not exceed the levels specified. Control at the occulus is effected by a large, circular sail fabricated of silicone-coated glass fibre. Using yachting technology for easy access for maintenance, the sail is stretched below the roof opening to provide approximately 50 per cent light transmission. Control at the clerestory windows is by electrically-controlled louvres which eliminate sunlight and allow sufficient daylight to enter the atrium, and add to the daylight available in the associated offices at the upper levels.

The artificial lighting is designed to complement the daylighting when necessary during the day, and provide a gradual transition from day to night when electric sources take over.

High pressure discharge sources are placed at high level: 400 watt metal halide downlighting is located on the service

Architect and Lighting Design: Arup Associates

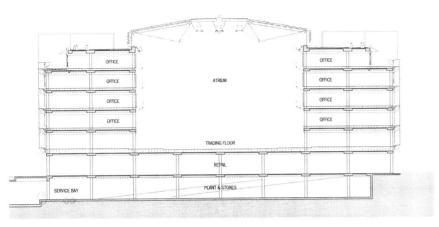

gantry above the central sail shading device, the fabric both diffusing the lights, and being itself lit by them. Uplighters from the service gantry around the perimeter light up the solid metal ceiling above, providing diffuse light to the space and reducing the visual contrast at ceiling level. Since this type of light source cannot be dimmed, it has been necessary to vary the level of light by switching, a third at a time to provide the necessary variation.

The lighting of the 18 metres of offices around the edge of the atrium is by recessed fluorescent fittings that are supplemented during the day by daylighting to reduce energy consumption. The building is air conditioned.

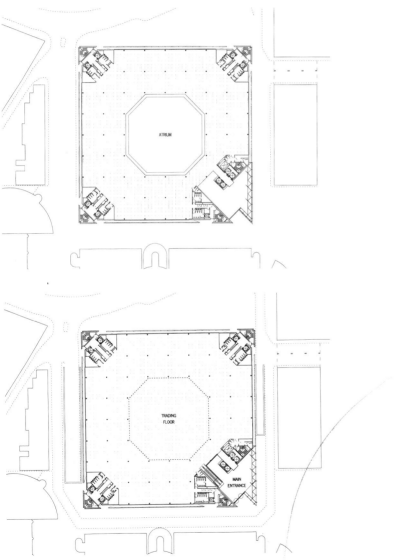

Sections through the atrium and offices.

General view of atrium (Photography Alan Williams).

Detail of the sail at roof level (Photography Alan Williams).

Section 4 Industrial

Case study 16 Dr Martens footwear factory, Wollaston

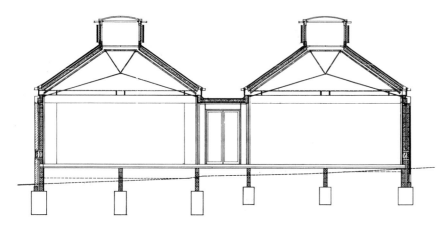

Cross-section to show the roof and clerestory lighting (Haworth Tompkins).

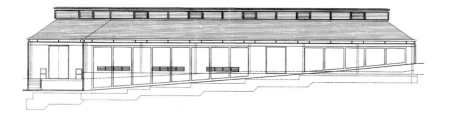

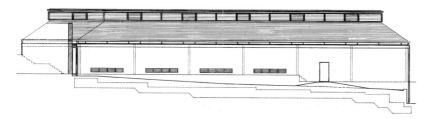

Long sections (Haworth Tompkins).

The comprehensive redevelopment of the company headquarters of the R. Griggs Group in Wollaston, the manufacturers of Dr Martens footwear, was designed by the architects Haworth Tompkins.

The brief to the architects was to provide an excellent working environment in a series of buildings which responded to the scale and grain of this Northampton village, maintaining the predominant palate of local building materials.

The buildings were visualized as contemporary barns, designed to accommodate the operational requirements of a factory and office building in a manner which would fit into the landscape and mature with age and use.

The choice of lighting strategy for the two pitched roof forms was developed to reinforce the appreciation of the buildings through the use of both natural and artificial light.

Natural light enters through continuous roof lights at the apex of the two pitched roof forms, supplemented by continuous clerestory windows at the junction between walls and roof. The system leaves the walls free of openings, to provide complete flexibility in developing present and future manufacturing processes.

The lanterns are roofed with insulated translucent plastic sheet, and faced with prefabricated aluminium insulated panels containing louvres at intervals. A steel walkway in the lanterns gives access to the louvre mechanisms and provides sun shading to the factory floor below.

The levels of natural light are supplemented by a series of perimeter tungsten halogen uplighters to reduce the contrast between the timber panelled ceiling and the roof lights.

The available natural light suffices for all but overcast days. The main artificial lighting derives from a specially developed fitting placed in the roof lights which takes over from daylight as light fades (see detail). The artificial light is designed to provide 300 lux to the work surface, but in addition task lights are available.

The combination of natural and artificial lighting systems provides the excellent environmental working conditions demanded by the brief.

Architect: Haworth Tompkins Architects;
Lighting Design: Jonathan Speirs & Associates

Exterior view of the factory (Photographer Anthony Oliver).

Interior of factory in the daytime (Photographer Anthony Oliver).

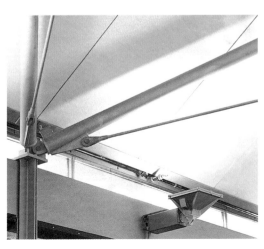

Detail of clerestory windows' structure and uplighter (Photographer Anthony Oliver).

Case study 17 Waste Disposal Plant, Tyseley

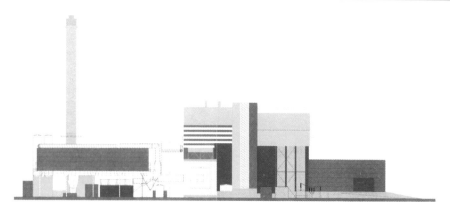

Elevation (Faulks Perry Culley & Rech).

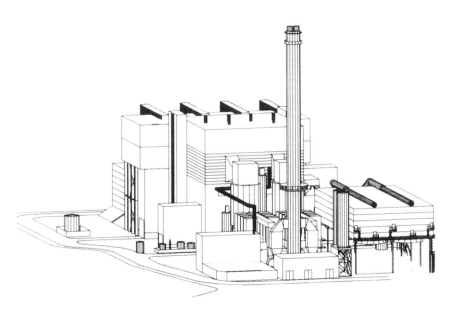

Axonometric of project (Faulks Perry Culley & Rech).

Completed exterior at night/blue (Photographer Anthony Oliver).

The building is a highly functional 'energy from waste' (EFW) plant by architects Faulks Perry Culley & Rech. The purpose of this case study, however, is to illustrate the night time appearance of the plant as devised by the artist Martin Richman.

There were four key factors in bringing this work to light:

1 The encouragement of Birmingham City Council, by planning guidance, suggesting 1 per cent of building costs be allocated for the commissioning of works of art.
2 The commitment of Tyseley Waste Disposal Ltd to the work.
3 The Public Art Commission Agency (PACA) proposal for the appointment of Martin Richman following a selection process.
4 The co-operation of the architects with the artist to ensure that the rigid configurations of the plant could be turned to positive advantage by the judicious use of colour and materials to achieve the planned lit appearance at night.

The function of the plant is to accept over 1000 tonnes of waste each day and to incinerate this so as to generate electricity.

The plant is automated. Although independent of the working of the plant the main cubic volumes are articulated with colour and light, whilst the movement of people and materials is suggested by interior animation, using translucent exterior elements glowing with light.

In the words of the artist, 'Lighting the outside of a building does more than just make the building visible in the darkness; new techniques use lighting to bring static objects alive . . . The heart of the building is a furnace, and I wished to make manifest externally some sense of its glowing core, an outward extension of its internal nature.'

This has been accomplished where possible by utilizing the interior functional lighting to redefine the structure with translucent panels that react to the changing light and colour within, giving a 'dematerializing' impression.

Richman uses the advanced technology

Architect: Faulks Perry Culley & Rech; Artist: Martin Richman

Completed exterior at night/red
(Photographer Anthony Oliver).

of the 'disco' (the DMX or digital multi-plexing controller) to achieve the movement and colour changes he has devised. The dichroic lights inside the box towers change from violet-blue to orange-magenta. Beams of light shoot vertically up the façade of the building twelve metres up to the towers above, using polycarbonate light guidance pipes.

Viewed at night from the nearby motorway the building is seen as a slowly changing lit sculpture and is clearly a local landmark.

Interior view towards
exterior/blue (Photographer
Anthony Oliver).

Interior view towards
exterior/red (Photographer
Anthony Oliver).

Case study 18 York Shipley factory, Basildon

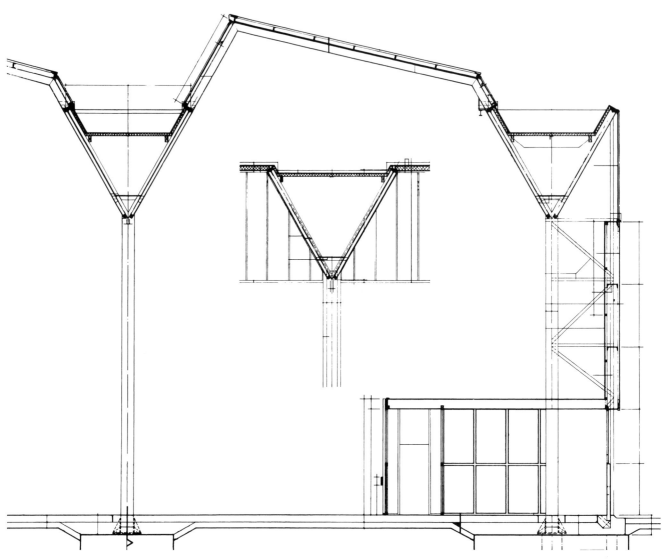

Section through the monitor roof and overhead duct (Arup Associates).

Despite work by the Building Research Station in the 1950s on roof monitors for the admission of daylight in factories, the early 1960s saw a trend towards the windowless factory. This was largely because the admission of sufficient daylight was obscured by the forest of pipes and other services hanging below.

The designers for the York Shipley factory went on to form the architects, Arup Associates. Using the BRS research, they created a modified monitor roof to satisfy a 10 per cent daylight factor. Through an integration of services at roof level allowance was made for window cleaning and the overhead maintenance of artificial light fittings. Other services

Architect and Services Engineers: Arup Associates

General view of the interior (Arup Associates).

requiring integration with the roof structure to avoid overhead obstruction were gas, compressed air, water and coolant, heating and ventilation.

The solution was an asymmetric east/west monitor where the south-facing windows were reduced to cut out solar gain, with larger windows to the north. The design was checked under the BRS artificial sky. The uncluttered roof provided the 10 per cent daylight factor and a triangular overhead duct between the monitors was provided for services.

For artificial lighting lines of fluorescent lamps were recessed into the sloping sides of the roof between the monitors, where re-lamping and cleaning could be carried out from inside the duct.

The result satisfied the client requirement for quality in the work environment with savings in energy, which at the time may not have been central to the architect's concept, but which in the light of present-day energy policy is clearly seen as far-sighted.

Case study 19 Brynmawr rubber factory, South Wales

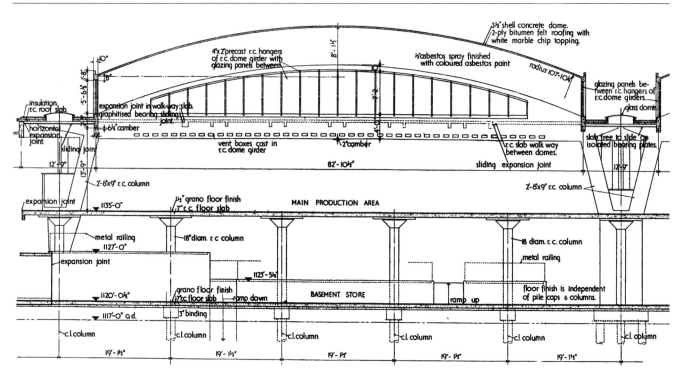

Section through dome (ACP).

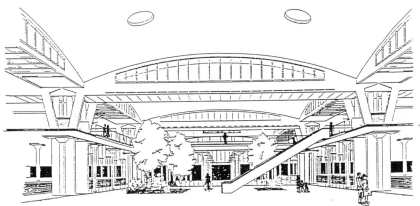

Architect's sketch for possible redevelopment (ACP).

Model of the whole complex showing the nine shell concrete domes (ACP).

The factory at Brynmawr in South Wales for Brimsdown Rubber Ltd, designed by the Architects Co-Partnership in 1945, was the first major post-war industrial building. It was completed in 1953 and in 1987 was described by Margaret Weston, Director of the Science Museum, as 'the most important post-war industrial building in Europe.'

Designed when steel and wood materials were in short supply, a decision was made to exploit the new form of shell concrete construction, which allowed maximum uninterrupted spans.

The roof form was designed to integrate both natural and artificial light. Similar circular openings were made in the shell concrete domes, some designed to admit daylight, others to hold fluorescent lamps which can be changed and serviced at roof level from above.

In addition, the daylight is supplemented by clerestory windows around the sides of each of the domes; and when daylight is absent, this is replaced by a continuous row of fluorescent lamps around the edges of each of the domes.

In the 1950s this lighting concept, where it is difficult to discern when

Architect: Architects Co-Partnership (ACP); Structural Engineers: Ove Arup & Partners; Lighting Design: Steenson Varming & Mulcahy

Daylight view of the interior (Copyright ACP).

daylight fades and artificial lighting takes over, was entirely new.

Sadly the building has fallen into disrepair, and there has been much discussion as to whether it can be refurbished and put to new use, such as a sports centre. It was Listed Grade II but the listing was removed in 1996 to make way for development, and today its future is in doubt.

Nightime view of the interior (Copyright ACP).

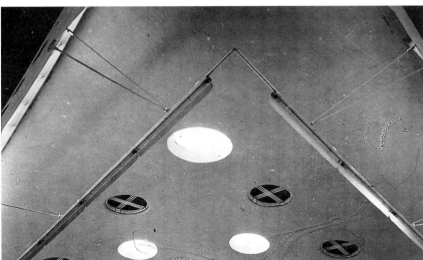

Roof light detail (Copyright ACP).

Case study 20 Boots factory, Nottingham

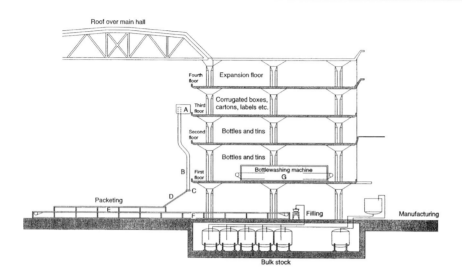

Cross-section through the building (derived from *The Bee* No. 13, Nov–Dec 1933).

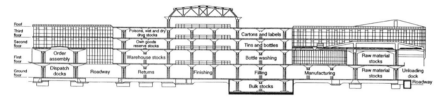

Long section through the building (derived from *The Bee* No. 12, April 1933).

Exterior of the building during the day (Copyright Boots).

Building of the Boots factory, designed by Sir Owen Williams, was started in 1927 when Lord Trent, the son of the founder of Boots, Jesse Boot, appointed the works planning committee to investigate the site in Beeston with a view to building the 'Ideal factory.'

When the factory opened in Beeston in 1933 in the middle of the depression, it provided the architect with the opportunity to apply fresh ideas and to use modern materials on a grand scale. Williams, with his engineering background, took on board the progressive ideas of the Bauhaus recently established in Germany, with its structural imperative and the use of reinforced concrete and glazed elevations.

The building was designed as a four-storey slab structure, based on a grid pattern of 2 metres, 34 centimetres. The floors were supported on columns with mushroom heads, leaving the sides of the building free of structure to allow floor-to-ceiling glazing. The building was ahead of its time in many ways including the provision of heating pipes placed along the sides, to act as both guard rails and to prevent condensation of the glass walls

The provision of good lighting was stated at the time to be of first importance and it was natural that the strategy for lighting would have been one of daylight, entering both through the floor-to-ceiling windows and through light wells created by puncturing the reinforced concrete roof above.

This roof was formed of small circular glass discs, 20 centimetres in diameter, inserted into the 5-centimetre reinforced concrete slab. One hundred and fifty thousand discs were employed in the roof construction, allowing a satisfactory level of daylight to infiltrate into the factory.

The artificial lighting was of a more primitive nature, reflecting the state-of-the-art of the day. A contemporary publication in 1935 stated 'The roof has been designed to admit the maximum amount of daylight, and carries high power bulbs which in winter flood the working tables with electric light.'

It should be remembered that fluorescent lighting had not been invented and filament lamps would have been the only

Architect: Sir Owen Williams

Exterior of the building at night (Copyright Boots).

practical type available. The amount of light derived from the method used would have been rather low. The lights are visible at high level in the picture and a more suitable method would have been to have had local sources close to the work.

Interior of the packing hall, 1933 (Copyright Boots).

Interior of the renovated hall with flying staircase, 1995 (Copyright Boots).

Section 5 Transport

Case study 21 Hong Kong International Airport

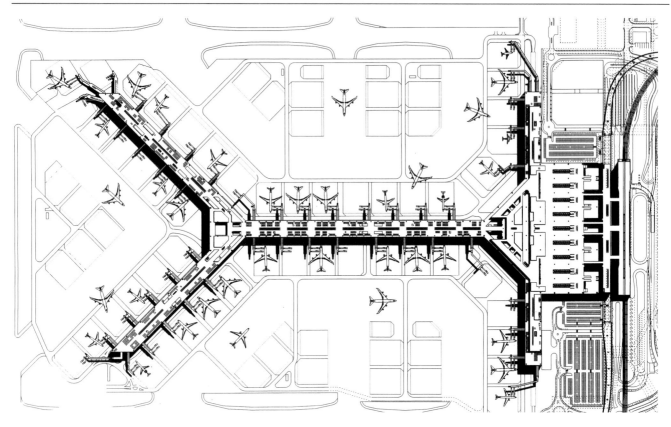

Ground plan of the departures concourse (Foster and Partners).

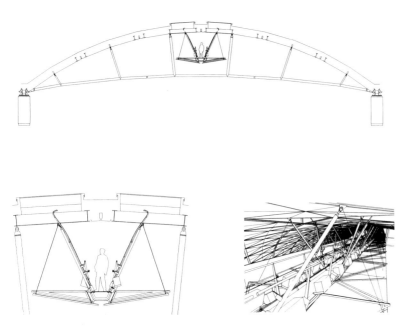

Ceiling gantry system to the departures concourse at first level (Foster and Partners).

The new international airport at Chek Lap Kok off Lantau island, Hong Kong, designed by Sir Norman Foster, is the largest airport in the world. It will be able to cope with 80 million passengers annually by the year 2040, equivalent to the current annual flow of Heathrow and JFK airports combined.

Air cooling in the hot climate of Hong Kong was a major consideration, and the air handling solution was part of an overall strategy to minimize heat gain and energy consumption. The lighting, both during the day and after dark, is an important aspect of this overall strategy.

Despite its deep plan (324 metres wide) the building is fully daylit from floor to ceiling through side windows and lines of triangular skylights in the centre of each of the nine barrel-vaulted roofs. The natural light is diffused by suspended reflectors along the roof gangway which provides access for maintenance, as well as support for the artificial lighting fittings. The skylights themselves are glazed with toughened laminated glass, coated to minimize solar gain; they cover some 6 per cent of the floor area of the terminal.

Architect: Foster and Partners;
Lighting Design: Fisher Marantz Renfro Stone

Departures check-in (Photographer Dennis Gilbert).

Departures spine during the day (Photographer Dennis Gilbert).

Arrivals meeting hall (Photographer Dennis Gilbert).

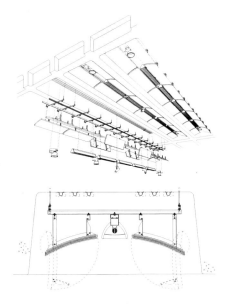

Detail of the special light fitting to indicate the acoustic reflector sides, removable for access and maintenance (Foster and Partners).

The artificial lighting in the concourse is provided by indirect uplighting, supported by the light reflectors' gantry. The light source selected is metal halide because of its 'colour rendering and colour temperature, efficiency and long life.'

In order to balance the daylight and at the same time reduce the use of energy, the lighting circuits are activated by a three phase operational control system which allows a balance to be struck with the level of daylight and creates a smooth transition between the two.

To avoid the impression of a uniformly-lit space and to add variety, certain areas are accentuated by additional downlights installed in such areas as atrium voids, immigration and check-in.

In the departures area the daylight is balanced by continuous wall washing, adding to the light on vertical surfaces and designed to give an uplift to the spaces.

In the arrivals hall a different lighting strategy is adopted, using lines of direct low brightness fluorescent fittings, direct mounted on to the concrete ceiling coffers.

Case study 22 Southampton International Airport, Eastleigh

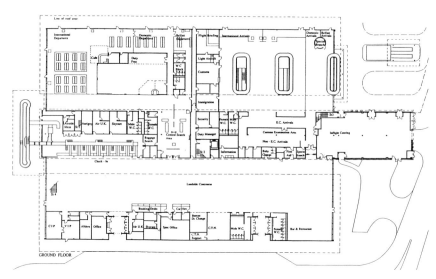

Plan of the airport (Manser Associates).

Southampton International Airport was designed by Manser Associates and completed in 1995. It is located at Eastleigh and was designed to serve the Southampton area and Portsmouth.

The lighting strategy is basically one of daylighting, and as can be seen in the associated section, daylight enters at roof level as well as from the sides, providing ample daylight to the booking hall.

Daylight is also the main light source for the waiting areas and cafeteria. A modest scheme of artificial lighting has been installed using pendant metal halide fittings and these provide the necessary light for the different spaces. They are similar to those used for the great London railway terminals' engine sheds.

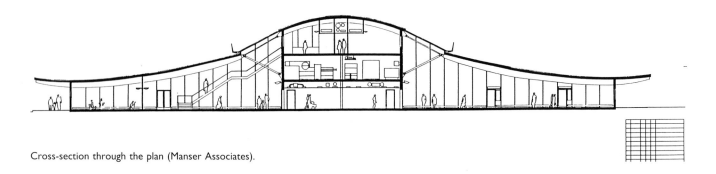

Cross-section through the plan (Manser Associates).

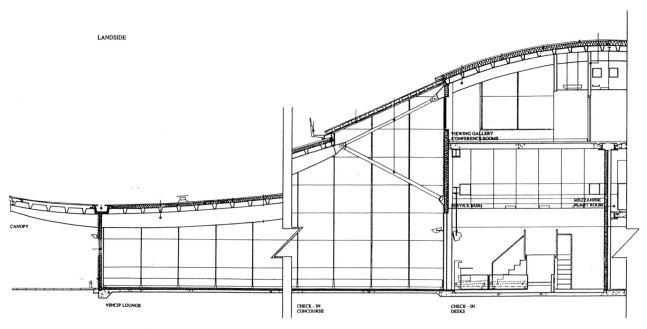

Detail of cross-section (Manser Associates).

Architect: Manser Associates; Lighting Design: DPA Lighting Consultants

Booking hall by daylight (Photographry Manser Associates).

Exterior view of airport at night (Photography Manser Associates).

Interior area by night (Photography Manser Associates).

Case study 23 Waterloo International Terminal, London

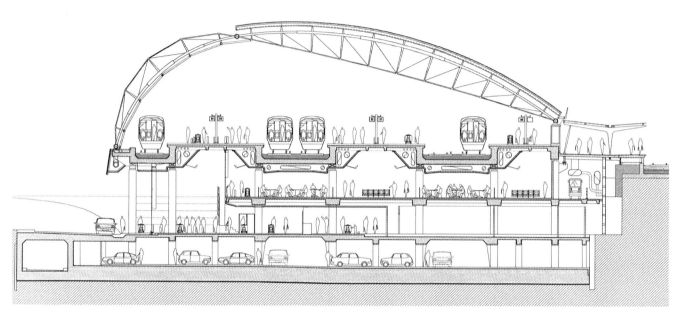

Section through arrival and departure halls (Nicholas Grimshaw & Partners Ltd).

Long section through the roof (Nicholas Grimshaw & Partners Ltd).

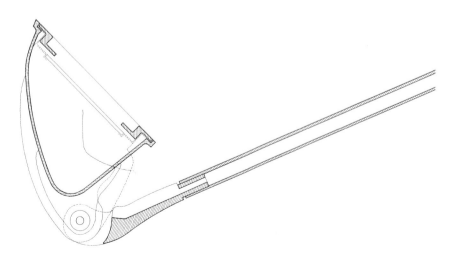

Details of light fitting (Nicholas Grimshaw & Partners Ltd).

The new terminal for Eurostar trains at Waterloo, designed by architects Nicholas Grimshaw & Partners Ltd, is in the great tradition of nineteenth century London railway termini. It was completed in 1993 over a century after engineers such as Brunel had created the large-span iron roofed railway stations, and Grimshaw has 'captured this heritage of heroic engineering, and created a terminal to evoke the spirit of the new age of European railways.' Despite this, in many respects the terminal echoes that of an international airport, being designed to handle fifteen million passengers a year.

The overall lighting strategy for the terminal is one of daylighting, with a linear glass roof structure a quarter of a mile long covering the entire platform area. It can be seen from the aerial view at night that the roof of the platform

Architect: Nicholas Grimshaw & Partners Ltd; Lighting Design: Lighting Design Partnership in association with Roger Preston & Partners

Aerial view at night of the roof structure (Photography Zumtobel).

area is formed of an elongated, curved glass spine from which transverse skylights run at right angles to ensure that the whole platform area is adequately lit by daylight during the day.

The artificial lighting of the platform area consists mainly of 1 kilowatt and 400 watt metal halide downlights hung from the roof structure with uplights incorporated only to light the ribbed steel roof sections in between the areas of glazing.

A glance at the cross-section indicates that daylight cannot fully permeate into the arrival and departure halls at lower levels, but at night the artificial lighting of both the platform area and departure hall below provide a unity of experience.

Underneath the main platform area the angled profile of the railway track beds forms the sloping ceiling to the departure hall. Here special light fittings have been designed to uplight the sloping ceilings to create a skylight effect.

These fittings were developed as 'specials' as a result of close co-ordination between the architects, lighting consultants and manufacturer. They consist of 150 watt metal halide uplighters with self-centring mechanisms, supported by 1-metre-long stainless steel tubes.

For the engine sheds/platform areas, one basic design was developed containing a variety of modifications to accommodate the changes of building section. The light is controlled by a parabolic reflector with integral black metal louvres to eliminate sideways glare, the lamp size being selected to cope with the light required in each location (see diagram).

The areas selected for this case study cover the more public aspects of the station, the platform and departure areas but the station also encompasses offices, a retail area, VIP lounges, services and car parking, and some 4500 lighting units were supplied to this vast complex. A large hidden area, referred to as 'the arches', has been refurbished to accommodate all the many services, from staff canteens to back-up power, quarantine and even prison cells.

Platform area during the day (Photographer Derek Phillips).

(Photography Zumtobel)

Section 6 Leisure

Case study 24 Sports hall, Bridgemary Community School, Hampshire

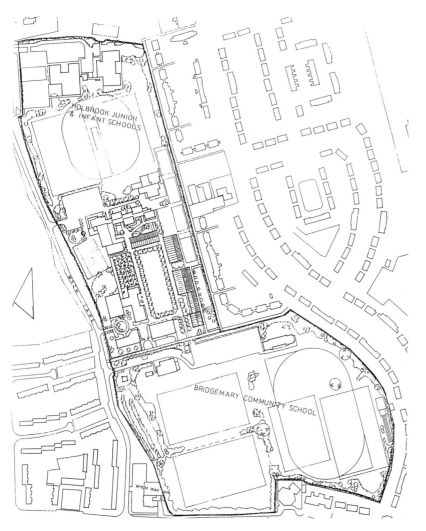

Site plan to show orientation (Hampshire County Council).

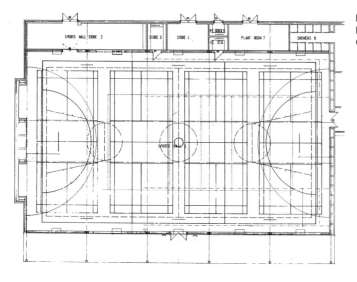

Plan of sports hall (Hampshire County Council).

In contrast to the workshop at Bisham Abbey (see Case Study 30), the sports hall at the Bridgemary Community School, designed by the Hampshire County Council Architect's Department and completed in 1990, adopts a daylight strategy for its design.

The daylight enters from a roof light extending the length of the building down the east elevation, whilst the pattern on the west side is purely decorative.

A sailcloth ceiling is stretched below the pitch of the roof to act as a luminous ceiling to conceal both the roof light itself and artificial sources placed above, as can be seen in the section on p. 151.

During the day the roof light gives diffused daylight over the sports hall and although it does not provide a view out, it does provide some information of the quality of the day outside.

At night artificial light takes over from the forty-five 250 watt high pressure mercury downlights located above the sailcloth. Upward light to the unlit area of the ceiling is provided by ten 250 watt metal halide uplighters with asymmetric reflectors mounted at the top of the brickwork side walls.

The lighting design illustrates the close co-ordination between the scheme for natural light and for artificial lighting, based upon low energy as a criterion. This makes an interesting contrast to the philosophy prevailing at the time of Bisham (1977) and the present day (see Case Study 30 Bisham Abbey Sports Centre).

Architect: Hampshire County Council Architects Department

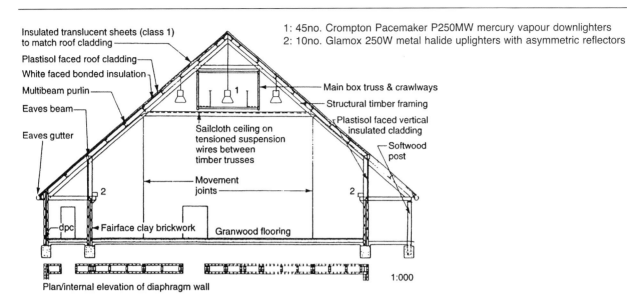

Insulated translucent sheets (class 1) to match roof cladding

Plastisol faced roof cladding

White faced bonded insulation

Multibeam purlin

Eaves beam

Eaves gutter

Sailcloth ceiling on tensioned suspension wires between timber trusses

Movement joints

dpc

Fairface clay brickwork

Granwood flooring

1: 45no. Crompton Pacemaker P250MW mercury vapour downlighters
2: 10no. Glamox 250W metal halide uplighters with asymmetric reflectors

Main box truss & crawlways

Structural timber framing

Plastisol faced vertical insulated cladding

Softwood post

Plan/internal elevation of diaphragm wall

1:000

Section to show daylight entry from the north east (Hampshire County Council).

Exterior seen from the north west end (Photographer R.A. Brooks).

Cross-view of hall (Photographer R.A. Brooks).

View down length of hall (Photographer R.A. Brooks).

Case study 25 Light Sculpture, Ilfracombe Pavilion

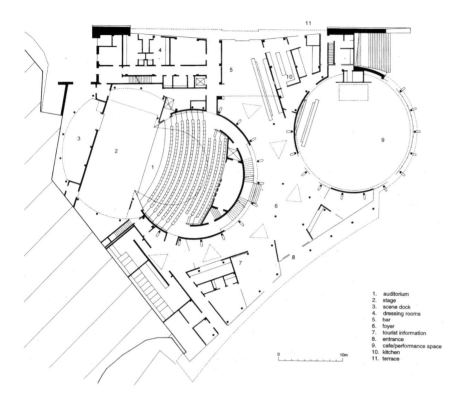

Plan of the overall pavilion (Tim Ronalds).

1. auditorium
2. stage
3. scene dock
4. dressing rooms
5. bar
6. foyer
7. tourist information
8. entrance
9. cafe/performance space
10. kitchen
11. terrace

0   10m

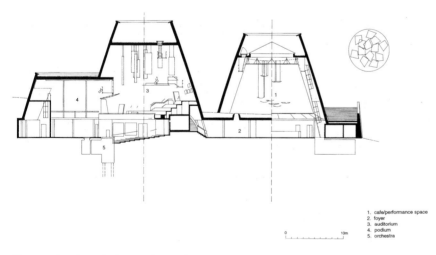

1. cafe/performance space
2. foyer
3. auditorium
4. podium
5. orchestra

0   10m

Cross-section through the complex to show the two cones (Tim Ronalds).

Set between the rocky expanse of Exmoor and the sea, on the North Devon coast, the town of Ilfracombe was developed as a Victorian seaside resort in 1840; but it was not until the town decided it should have a 'stately pleasure dome' that the Ilfracombe Pavilion was born. The building, designed by architect Tim Ronalds, consists essentially of two large cones set into a landscaped foyer at ground level, the one containing the Landmark Theatre and the other the Winter garden.

The Landmark is a 500 seat auditorium used for stage and cinema, set on two levels with no natural light, being lit by normal theatre lighting. The Winter garden on the other hand is flooded with daylight during the day from the large occulus at the top of the cone, and it is this that forms the basis for this case study.

The Winter garden has been a traditional community space in our seaside resorts from the Victorian age. Here it is used as a café during the day and for cabaret, dances and concerts in the afternoon and evening. The space can also be used for exhibitions and private social events, a tremendous asset to the facilities of a seaside resort.

The lighting concept for the space is of particular interest, based on the idea of a suspended white cone representing the changing sky over the town. The lighting consultant conceived that red, blue and green light derived from tungsten halogen projectors with coloured filters could wash the cone, leaving the suspended plywood acoustic panels in silhouette; and that an associated dimming system could provide a multitude of colours, with eight changing sky scenes.

A sophisticated control system was programmed to cross fade to provide a series of gradually changing linked scenes over a ten minute cycle. The cycle for the light sculpture could be manually programmed to provide single effects of full colour or more subtle shades, or be activated by photocell and time clock.

Since the effect of the system really comes into its own after dark, and in order to save energy, it is operated only when daylight fades, and then underrun by 10 per cent to save energy.

Architect: Tim Ronalds Architects; Services Engineer: Max Fordham & Partners; Specialist Lighting Design: Speirs & Major

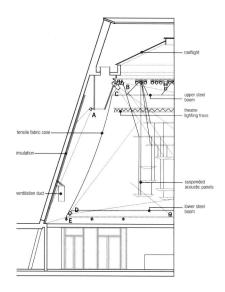

Cross-section through the Winter garden (Tim Ronalds).

Interior of the Winter garden (Photographer Martin Charles).

Views of the light sculpture (Photographer Mark Major).

Case study 26 Bar 38, Manchester

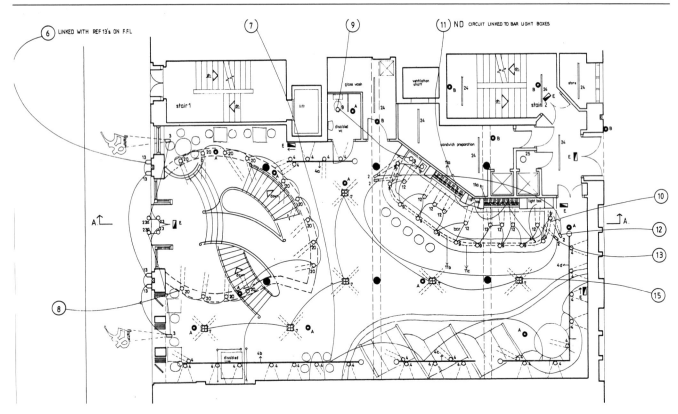

Plan of ground floor, showing the lighting layout (DPA).

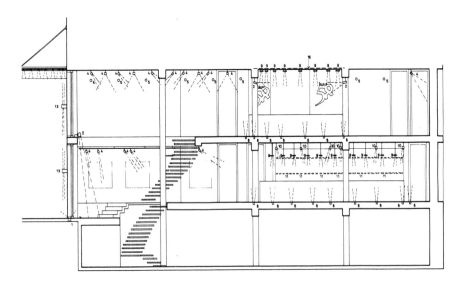

Section to show the staircase and circulation (DPA).

The lighting strategy for the Bar 38 in Manchester is essentially one of artificial light, both during the day and at night. The aim is to provide pools of local light using mainly low voltage spots mounted on track, directly on to the ceilings or in some places, like the bar, recessed to define its curved shape.

Because the low voltage spots are adjustable, they can be focused on to the tables or the works of art on the walls, allowing for a flexibility to adapt to changes in the decorations and to current fashion.

The bar itself is outlined with blue, cold cathode lamps which take up the configuration of the curved front. The staircase, which is a dominant element of the design, giving access to the three levels of the bar, has a special suspended array of coloured glasses taking up its curvilinear shape. This is engineered by using a single 150 watt metal halide lamp coupled to a series of fibre optic outlets, one to each of the coloured glass elements. This has the advantage of ease of maintenance in that only one lamp is employed in a location where it can be replaced easily. The aim of the bar

Interior Designer: Newman Carty & Gauge;
Lighting Design: DPA Lighting Consultants

View of the bar (Photographer Derek Phillips).

restaurant is to be dramatic, and to provide customers with their own individual environment.

Daylight has not been entirely forgotten, since floor-to-ceiling windows at the front allow natural light to enter for the first four-metre-depth at ground and first levels (see section). During the day this gives a specific quality to the interior, and the tendency on fine days is for customers to find tables close to the front. The weather in Manchester being what it is, fine days are all too few, so the lighting strategy for the bar is for artificial lighting to predominate, both during the day and at night; a solution dictated by the site conditions, where the building is located between two buildings, daylight entering only from the front.

Decorative lighting to staircase (Photographer Derek Phillips).

General view of restaurant (Photographer Derek Phillips).

Case study 27 Inland Revenue, Nottingham

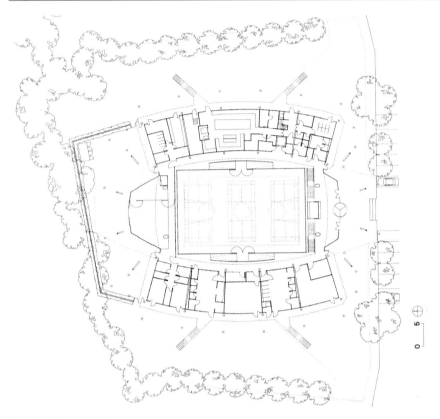

Plan (Sir Michael Hopkins).

The new Amenity Building is a part of a large group of offices built to accommodate the staff of the Inland Revenue who relocated to Nottingham.

The brief provided for the whole site was to design an energy-efficient building complex; one in which air conditioning would not be required, and where natural daylight and ventilation were to be applied.

The purpose of this area of the complex was to provide social amenity facilities, including a large sports hall, gym and staff restaurant.

The Amenity Building displays a daylight strategy, being formed of a tented structure through which shaded natural light enters and where artificial light is not required during the day. The surrounding areas of the staff restaurant are also provided with ample daylight.

In the surrounding lowered ceiling areas recessed filament lamps are set into the ceiling, whilst spotlighting is derived from fittings placed around the main structural masts at a high enough level to eliminate any glare to the restaurant area. This ring of fittings is designed to light both downwards to the play area and upwards to lighten the roof.

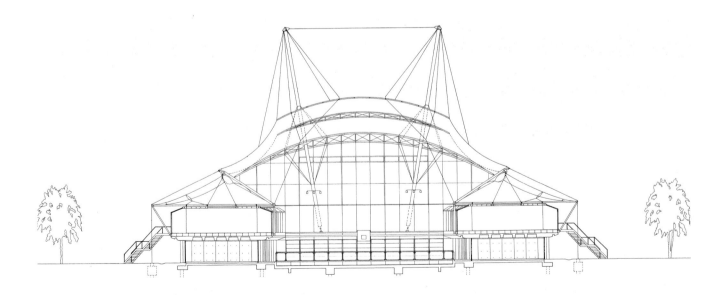

Cross-section (Sir Michael Hopkins).

Architect: Sir Michael Hopkins

External view by day (Photographer Derek Phillips).

Side view over-looking sports hall (Photographer Derek Phillips).

The sports hall (Photographer Derek Phillips).

Case study 28 Opera house, Helsinki

Exterior view by day (Erco Lighting).

Exterior view at night (Erco Lighting).

The foyer by day (Erco Lighting).

The building for the Finland National Opera Company was the result of an architectural competition won by Finnish architects Eero Hyvamaki, Jukka Karhunen and Risto Parkkinen in 1977, and followed the building of Finland's first concert hall, the Finlandia Hall, designed by Alva Aalto in 1971.

The opera house is a good example of daylighting where it is appropriate in the foyer, offices and restaurant encouraging views out to the countryside and lake beyond, whilst the interior of the opera house itself is lit by normal artificial theatre lighting. The natural light for the staircases is replaced at night with tungsten halogen floodlights, to create a similar spatial effect.

The main auditorium is lit using tungsten halogen downlights, doubled up to provide highly-illuminated areas when the auditorium is not being used for performances. The parapets use cove lighting.

Accent lighting in the auditorium and restaurants uses low voltage, directional downlights, with some added pendant lighting. Using tungsten and tungsten halogen lamps, all the lighting is capable of being dimmed to extinction when required.

Seen from outside at night, the opera house demonstrates the way in which the internal structure is what is experienced through the transparent glass enclosure, and the importance which should be given to ensuring a credible structure.

Architects: Eero Hyvamaki, Jukka Karhunen & Risto Parkinen; Lighting Consultant: Joel Majurinin Ky

The foyer at night (Erco Lighting).

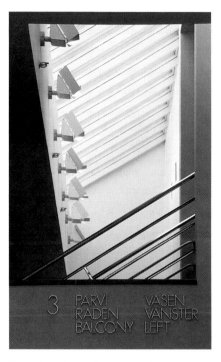

View of uplighters (Erco Lighting).

The restaurant (Erco Lighting).

Case study 29 Copley Restaurant, Halifax Building Society HQ

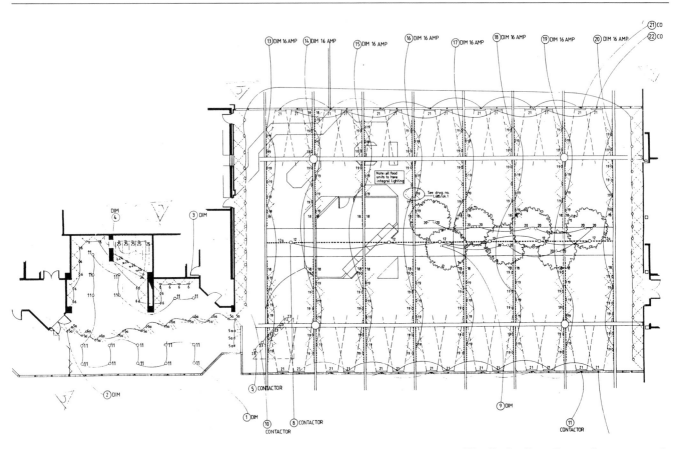

Plan of the restaurant, with lighting plan (DPA).

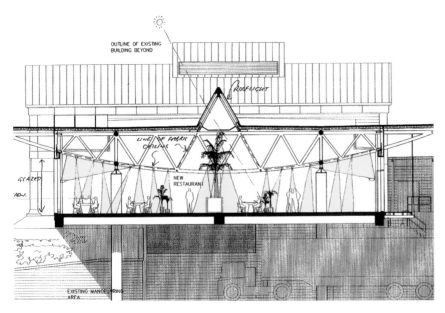

Schematic section to show lighting (DPA).

The Copley Data Centre forms a part of the Halifax Building Society Headquarters complex and the restaurant, designed by architects Abbey Hanson Rowe and completed in 1996, has been built to serve 1800 staff.

The staff using the restaurant spend much of their day working at VDUs, and the lighting quality for the restaurant was considered to be of prime importance.

The following simple criteria were adopted:

1 Maximum use to be made of daylight, with dynamic views out to the exterior landscaping. Naturally lit space to provide a dramatic contrast with the main office's fluorescent lighting to assist with relaxation.
2 A variety of internal spaces to be provided, to include a restaurant, coffee bar and lounge or sitting area, each with their own characteristic atmosphere, but linked together as a sequential experience.

Daylight is provided in the restaurant by four-metre high glazed walls along the

Architect: Abbey Hanson Rowe (AHR);
Lighting Design: DPA Lighting Consultants

General view of restaurant (Photography AHR).

two main elevations, associated with a central roof light. The walls provide the views out as well as high quality daylight, and this coupled with the central roof light, which extends for the whole length of the space, provide the whole restaurant with natural light.

The structure for the roof consists of a series of curved roof trusses, and these are used as the support for fabric sails stretched between; these are used to lower the scale, giving a visual ceiling.

Below the central roof light the sails are omitted to provide a sense of upward relief. Underneath, a series of large plants are located, helping to divide up the space. The planters incorporate upward floods to light the leaves from below, casting changing shadows on the ceiling.

The artificial lighting is designed to complement the architecture and emphasize the structure rather than the fittings themselves; the sails are uplit from 150 watt wall-mounted floods, and the restaurant tables are spot-lit by 50 watt low voltage lamps mounted on track fixed to the trusses between the sails (see detailed photograph).

The artificial lighting for each of the three spaces gives emphasis and variety to suit their individual function. The other areas the coffee bar and the lounge are individually treated.

The whole complex uses tungsten filament sources, underrun to improve their life and reduce the energy used. The lighting control is by scene sets dimmer systems, allowing a variety of effect as daylight fades, the stimulation of daylight changing to a more relaxing atmosphere after dark.

Restaurant by day (Photography DPA).

Restaurant by night (Photography DPA).

Case study 30 Bisham Abbey Sports Centre, Marlow

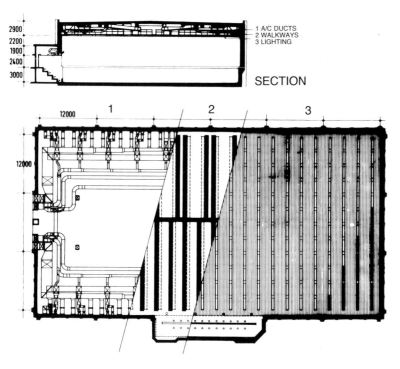

Plan and section of the workshop (DPA).

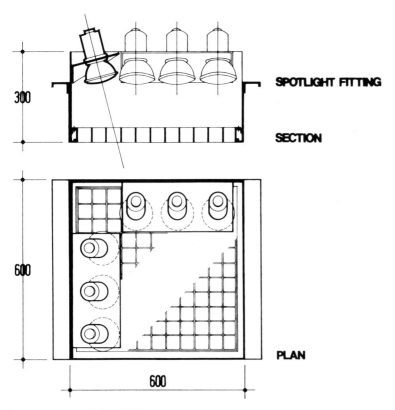

Box containing spotlights (DPA).

Bisham Abbey, in its attractive setting on the banks of the Thames near Marlow, dates back to the twelfth century. Since 1946 it has been administered as a National Sports Centre. Whilst the centre offered outdoor facilities, by 1973 it was decided to build an indoor facility for a variety of sports together with hostel accommodation for visitors and teams.

The Sports Council's brief to the architects, Rice/Roberts, was that they were to create a centre of excellence for the training of national teams, and whilst the centre provides for other sports, the main element of accommodation is planned as a sports workshop, which at 72 metres by 36 metres was large enough for training in the fields of football, rugby, hockey and tennis.

The consultants' early studies suggested that some form of limited daylight entry to the workshop would be desirable, but this was ruled out on the grounds that it would have made it more difficult to guarantee consistent environmental conditions. Instead, a decision was made to create the building as a blind box that eliminated all natural light; daylight was encouraged in other areas of the building.

The environmental conditions requested for the workshop were as follows:

Temperature not to exceed 13 degrees centigrade
Relative Humidity 50 per cent
Noise Rating of NR40

Since the workshop would be lit by artificial lighting at all times when in use, and because it was far from certain what the ideal light for each sport might be, a degree of variety or flexibility was to be made possible by suitable switching arrangements. The variety made available consisted of three levels of illuminance, 900, 750 and 600 lux. The light came from lines of recessed fluorescent fittings each containing three lines of lamps and the fittings were designed to be serviced from the ceiling space above, which was required for the air conditioning ducting

Architects: Rice/Roberts; Lighting Design: DPA Lighting Consultants,
in association with Robert Somerville Associates, Consulting Engineers

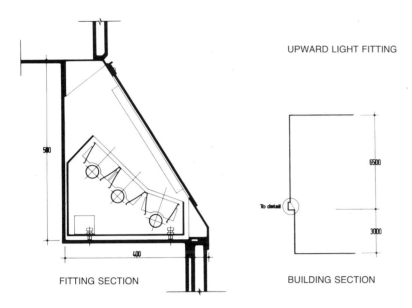

UPWARD LIGHT FITTING

FITTING SECTION   BUILDING SECTION

Uplight (DPA).

Exterior of sports hall workshop (Photography Derek Phillips).

Interior of sports hall (Photography Derek Phillips).

(see Plan of Service ceiling with associated section).

Between each fluorescent fitting was placed a box containing directional filament spots to enable a balance to be obtained between the fluorescent and the directional light components, increasing the flexibility of solution available.

Other aspects of the design included lines of low level upward light to reduce the contrast between the brightness of the lines of fluorescent light and the darker ceiling between; good colour rendering; the lowest heat load consistent with the stated objectives; and economic installation and overhead maintenance.

Constraints imposed upon the design included the need to protect the upper and lower light fittings against impact from missiles and control of the relationship of airflow through the fluorescent fittings with the air conditioning ducting.

There is a viewing gallery down one side of the workshop where the environmental conditions needed are of a less stringent nature. One important requirement is that the void of the gallery should not appear as a black hole from the workshop itself, which is achieved through lighting added to the rear wall of the gallery behind the viewers.

Whilst lighting technology has advanced in terms of the colour and efficiency of the lamps available, and most particularly in the field of light controls, bearing in mind the initial decision to eliminate daylight, the solution adopted was state-of-the-art in the early 1970s. It was a typical example of the low importance attached to daylighting at the time and it is unlikely that a blind box facility would be built today to serve the same needs (see Case Study 24 Sports hall, Bridgemary Community School).

Case study 31 Swimming pool, Haileybury School, Hertfordshire

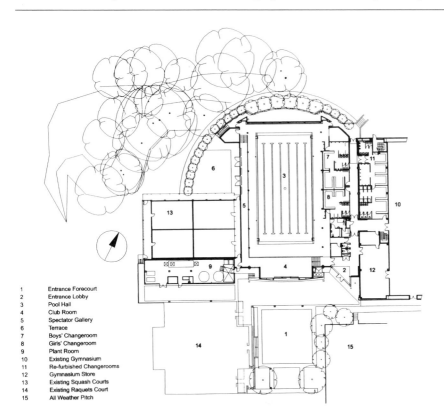

1   Entrance Forecourt
2   Entrance Lobby
3   Pool Hall
4   Club Room
5   Spectator Gallery
6   Terrace
7   Boys' Changeroom
8   Girls' Changeroom
9   Plant Room
10   Existing Gymnasium
11   Re-furbished Changerooms
12   Gymnasium Store
13   Existing Squash Courts
14   Existing Raquets Court
15   All Weather Pitch

Plan of the building showing orientation (Studio E).

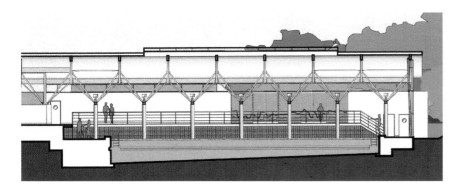

Long section (Studio E).

When the architects, Studio E, were asked to design a covered swimming pool (in 1996) for one of England's foremost public schools (a school built in 1806 by William Wilkins, architect for the National Gallery, to train the young gentlemen of the East India Company) it might have been expected that a rather traditional structure would have emerged but this was not the case.

The total funding for the project came from one of the Old Boys, but this did not encourage extravagance, and the overall costs for the project were strictly monitored to ensure value for money resulting in a total cost of less than that for a basic swimming pool providing equivalent facilities.

The key elements of the architects' design were, a highly sculptural floating roof supported by structural trees, associated with glazed walls permitting views out to the countryside. A cross-section was developed (on p. 165) to illustrate the floating nature of the roof.

The lighting design was based on a low energy policy, whereby the pool is lit during the day entirely by natural light entering from the glazed end walls and a central roof light over the pool. At night uplight from metal halide floods located at the intersections of the structural trees provides sufficient light downwards through reflection from the metal ceiling. Both systems eliminate the problem of glare from the water, which, if uncontrolled could present a problem of safety; the lighting of swimming pools requires care to eliminate the danger of surface reflection off the water as it prevents a life guard from seeing to the bottom.

The north of the pool faces mature woodland, and this has been exploited in the design which provides extensive glazing facing this direction which has none of the problems associated with sunlight and glare. After dark gentle flood lights illuminate the perimeter woodland adding drama and extending the impression of space within the building.

The perceived quality of the space is external and gives a bright daylit impression even on dull days. The controlled use of blues and white colours combines with the natural colours of the woodland to provide a natural ambience for the space.

Architect: Studio E Architects;
Services Engineers & Lighting Design: Max Fordham & Partners

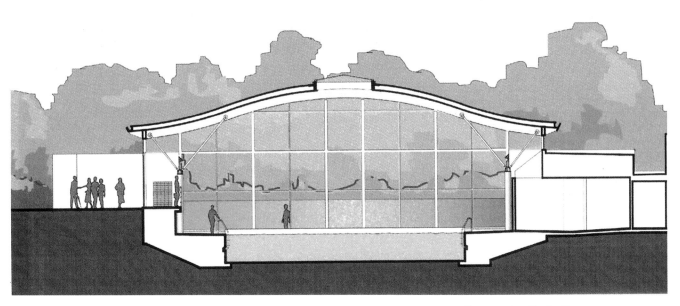

Cross-section (Studio E).

General interior view of pool (photographer Derek Phillips).

Detail of upward lights at the junction to the columns (photographer Derek Phillips).

Case study 32 Hilton Hotel, Heathrow (originally the Sterling Hotel)

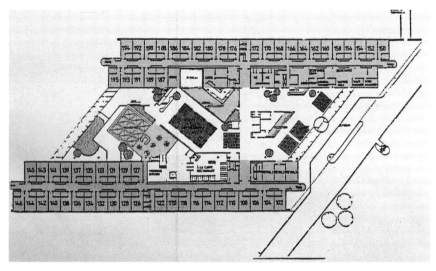

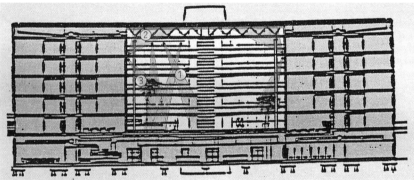

Plan and section (Derek Phillips).

Hotel exterior (photographer Derek Phillips).

Completed at the end of 1990, the Sterling Hotel at Heathrow's Terminal 4 was one of three British Airport Authority (BAA) hotels opening at the major British airports of Heathrow, Gatwick and Stansted. The Sterling Hotel at Heathrow (which was subsequently renamed the Hilton) was designed by Manser Associates and manages to capture the spirit of air travel and airport construction as no other.

The parallelogram plan, flanked by bedrooms down each side, allowed a large uninterrupted atrium space between, where all the major facilities of foyer, bar, brasserie and swimming pool are planned. The atrium relies principally on daylight during the day gained from the large glazed areas at each end of the building, together with three natural lighting slots at roof level. The natural light is supplemented on dull days by enhanced daylight from artificial sources controlled by computer, placed at high level in areas of the plan requiring emphasis, such as the sitting area close to the reception, and the brasserie.

A complete change of atmosphere at night is provided by artificial sources placed at low level, emphasizing the different elements of the hotel plan. Here the atrium appears as a series of lit focal areas. To ensure that the tall ceiling to the atrium is not dark at night it is flooded with blue light from the sides, which is maintained from the cleaning catwalks. Upward light has been built into the base of the real trees for night time lighting, but to ensure growth of the trees these are lit from powerful long life, low energy floodlights at high level during the day.

An important element of the architect's design is the stack of glass lifts connected by balconies which provide access to the banks of bedrooms on each side. Miniature spots are built into the bridges to light up to the ceiling (the underside of the bridge above) and give a pattern of light on the bridges as seen from below.

The combined strategy in this building of daylight during the day, supplemented by enhanced daylight from artificial

Architect: Manser Associates; Lighting: DPA Lighting Consultants

Hotel interior at night (Photographer Derek Phillips).

sources only when daylight is insufficient, differs from that adopted in most hotels where artificial lighting is used at all times. The architect, together with the interior designer Peter Glynn-Smith, thought of the atrium as a street open to the sky, a Champs Elysée in which all the public elements of a modern hotel could exist, and for the most part this has been achieved.

Swimming pool at night (Photographer Derek Phillips).

The Grill at night (Photographer Derek Phillips).

Case study 33 State and University Library, Göttingen

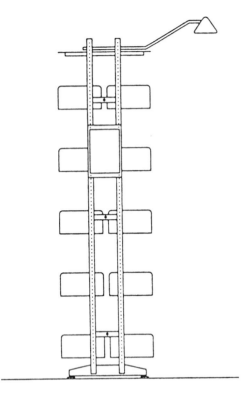

The lighting of the book stacks (Erco Lighting).

General view of reading room (Photography Erco Lighting).

Daylighting of the reading room (Photography Erco Lighting).

The State and University Library in Göttingen is one of the most important sources of literature in Germany, holding 3.6 million volumes, and is one of the country's five largest libraries. The building was the result of an architectural competition, won by the architect Professor Gerber in 1985.

The overall facility is organized into four primary functional areas:

Reading and Borrowing area;
Administration;
Storage and technical services;
Underground garage and building equipment.

A central determinant of the architecture, and an important contribution to the daylight entering from side windows, was the use of roof lighting, arranged lengthways on the finger-shaped building elements. The natural light is supplemented by artificial light adapted to the individual requirements of each space.

After dark, pairs of 250 watt broad and narrow beam tungsten halogen reflector lamps are trained through the transparent glass roof lights to bring artificial light into the halls.

The shelving is a dominant element in the reading rooms. It is lit by linear profile fittings using a 36 watt fluorescent lamp projecting forward from the top of the book stacks and uniformly lighting the vertical surfaces of the books whilst providing adequate light for visitors standing reading at the stacks.

In a similar way, the catalogue cupboards also have the projecting linear lights, but these are fitted with dark-light reflectors, so that the desks can be used as EDP stations, the fittings appearing dark even when lit from normal angles.

Beyond the book stacks, the sitting areas in the reading rooms rely on daylight and view during the day, monitored for sun and sky glare by external textile sunshades. Work places in the reading rooms have individual reading lamps at night, fitted with compact fluorescent lamps. The staff rooms are all daylit during the day.

Whilst the internal areas of the book

Architect: Prof. Gerber Partners; Lighting: Atelier für Lichtplanung Kress & Adams, Cologne in association with Erco Lighting

Book stack lighting, detail (Photography Erco Lighting).

stacks require artificial light during the day, the general impression of the space and particularly the reading areas is one of a well daylit space; an example of a successful daylighting strategy in a low energy environment.

Lighting the book stacks (Photography Erco Lighting).

Case study 34 Eton College Drawing School

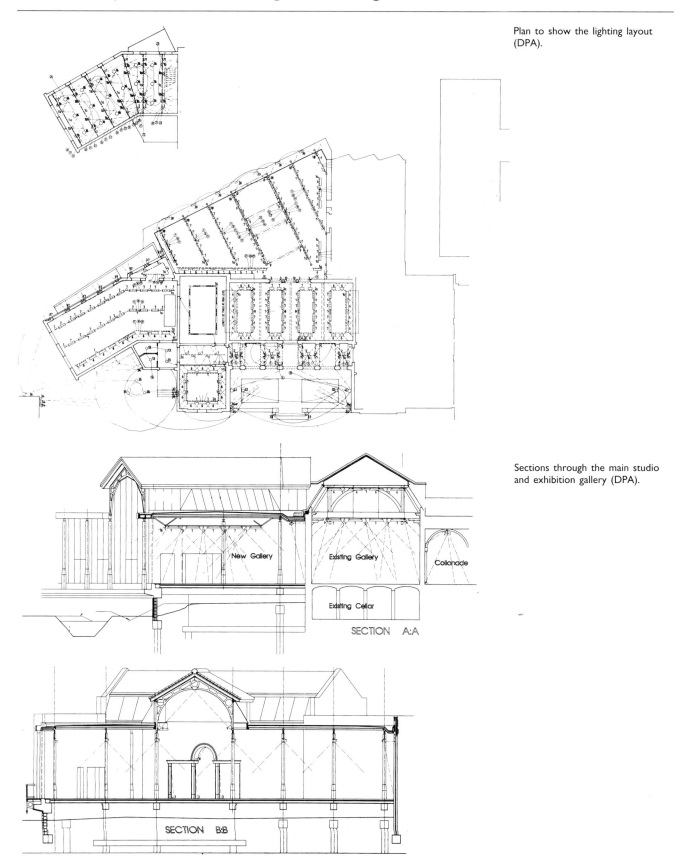

Plan to show the lighting layout (DPA).

Sections through the main studio and exhibition gallery (DPA).

New Gallery

Existing Gallery

Colonnade

Existing Cellar

SECTION A:A

SECTION B:B

Architect: Herbert J Stribling;
Lighting Design: DPA Lighting Consultants

First floor studio with side and overhead daylight (Photographer Derek Phillips).

Exhibition gallery top lit by day (Photographer Derek Phillips).

Stage 1 of the Drawing School, designed to be built in two stages, was completed in 1996. The daylighting to the drawing school has been designed by architect Herbert J. Stribling to provide natural light from the north through wall-to-ceiling glazing to the large studio at ground level. A north light roof at first floor level is associated with side windows, ensuring that light suitable for painting, drawing and sculpture is available at both levels. To the south the façade is more enclosed to the entrance colonnade, leading into the existing top-lit exhibition gallery.

The brief to the lighting consultant was to design an artificial lighting system that would provide a 'flexibility of approach,' with different lighting scenes to meet the needs of the school for different activities and as times change. The school is heavily used by the pupils at night, both in the winter and in the summer, so the lighting strategy both during the day and at night was most important.

The lighting method chosen was track-mounted at each ceiling truss. Lines of fluorescent fittings were interspersed with narrow beam spots which can be controlled to pick out special features.

The views out from the large, glazed north wall over a small river towards the playing fields are an inspiration and a fitting background to the changing environment of the studio interior, as is the view into the building seen at night.

View from exhibition to main gallery at night (Photographer Derek Phillips).

Case study 35 Cranfield University Library, Bedford

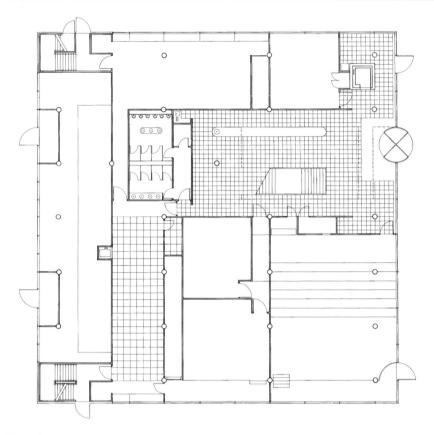

Floor plan at ground level (Foster and Partners).

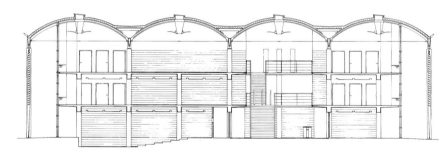

Section (Foster and Partners).

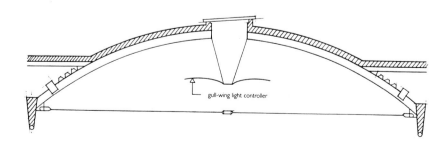

gull-wing light controller

Detail of the gull-wing light fitting (Foster and Partners).

The library at Cranfield University, designed by architect Sir Norman Foster, is intended to provide a much-needed hub for the university, while at the same time revising the concepts for a library in the information age. Cranfield is predominantly a post-graduate facility with expertise in engineering, business management and research; students have on-line connections to computer networks and electronic data bases.

It is still necessary to have book stacks, and seven kilometres of open shelving are retained at the top two floor levels, freeing up the ground floor for closed access archives and for social use, including a coffee bar and conference facilities.

The building is made up of a series of barrel vaults, and as can be seen in the cross-section, there is a daylight slot at the top of each barrel, with daylight also entering from large side windows. Students' desks are located along the sides of the building and sky glare is controlled by a series of external horizontal louvres which also permit views through to the exterior.

The artificial light comes from lines of fluorescent fittings suspended below the lighting slots, designed to provide the necessary downlight at night, but also giving reflected light up to the underside of the barrels during the day. Even when the artificial light is used the impression is still one of daylit space (see detail of the gull-wing light fitting. The detail, whilst not the same, derives from the same philosophy as the architect's work at Stansted Airport). The book stacks themselves are lit from recessed fluorescent fittings placed in the lowered ceilings of the first floor areas.

Architect: Foster and Partners;
Lighting Design: George Sexton in association with J. Roger Preston

Exterior night view (Photographer Dennis Gilbert).

Side view to show student desks and sun control louvres (Photographer James Morris).

Interior of atrium (Photographer Derek Phillips).

Case study 36 The Aldrich Library, University of Brighton

Section through building (Long & Kentish Architects).

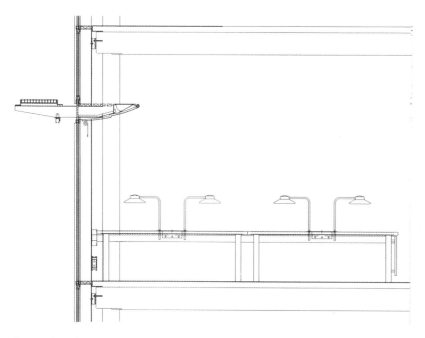

Section through exterior wall to show light shelf and table lights (Long & Kentish Architects).

The Aldrich Library at the University of Brighton, designed by the architects Long & Kentish, resulted from a competition won by the architects in 1994, the building being completed for occupation in 1996.

An early decision was made not to air condition the building, but to provide comfortable environmental conditions, where the reading areas were naturally lit, but the book stacks themselves were lit artificially by specially-designed lines of pendant fluorescent light fittings parallel to the stacks.

The library users could read by natural light on normal days, but were provided with low-level task lights which could be operated at their own control where daylight was thought to be insufficient. Each reading position was supplied with all the necessary servicing to enable the use of laptop computers and giving access to the internet. The building is equipped with 130 computer work stations, and houses 200 000 books and journals.

The positive reading environment and the expenditure of limited financial resources on electrical and electronic servicing rather than air conditioning proved to be the right decision, providing the right working conditions while reducing the electrical energy requirements to an acceptable degree.

The low-energy environment demanded a low-cost, low-energy mechanical heating and ventilation system. A mechanical extract system draws outside air through the façade, using a combination of a light shelf and an acoustically-lined air inlet (this can be seen at each floor level in pictures of the façade of the building). The system is combined with simple perimeter heating.

Architect and specialist Lighting Design: Long & Kentish Architects; Structural and Services Engineers: Ove Arup and Partners

Detail of sun control louvres (Photography Long & Kentish Architects).

View of library at night (Photography Long & Kentish Architects).

View to show the individual task lights (Photographer Duncan McNeil).

Case study 37 Lycée Albert Camus, Fréjus, Côte d'Azur

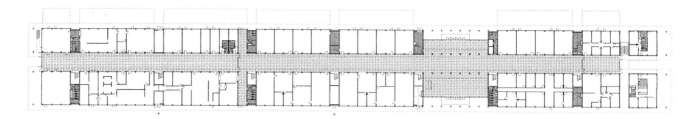

Ground plan (Foster and Partners).

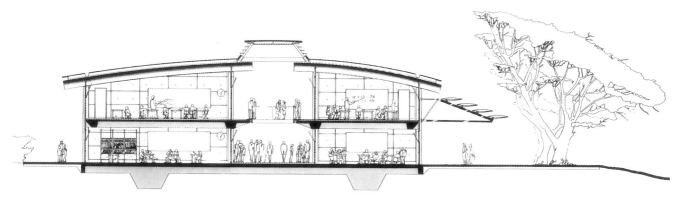

Cross-section (Foster and Partners).

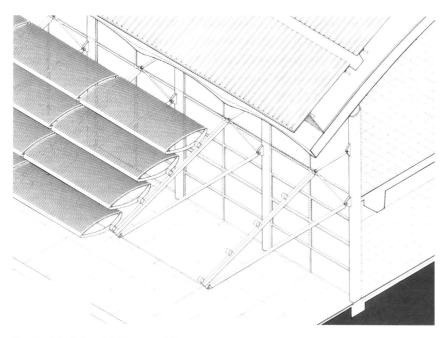

Detail of the *brise soleil* (Foster and Partners).

The Lycée Albert Camus in Fréjus on the Côte d'Azur is a new type of French school designed to offer 'partly vocational training' to children in the final three years of schooling. Designed by Sir Norman Foster, the school is based on a linear plan in response to the site, the social structure of the school and a low-energy concept that reacts to the Mediterranean climate.

A linear street forms the spine joining all the accommodation. The roof to the main two-storey building is formed from barrel vaults which meet at right angles to the street. The central spine, or street, is bisected by the entrance space which like a Mediterranean village square is a meeting point for the students, with a café and casual seating. It separates the public side on the north from the private side with its superb views to the south.

The daylight entering the two long sides is controlled to the south by overhanging *brise soleil* which provide a band of dappled shade from the overhead sunlight.

The aim of this 'passive' building is to reduce the necessity for 'active' environmental services to a minimum and to achieve a low-energy solution, the form and construction of the building playing a

Architect: Foster and Partners;
Lighting Design: J. Roger Preston & Partners

Exterior, north side (Photographer Dennis Gilbert).

Exterior to south, showing dappled shadows (Photographer Dennis Gilbert).

major part in the heating, cooling and ventilation.

The building clearly adopts a daylight strategy that reacts to the climatic changes outside. Light is received from a linear roof light extending the length of the building above the central double-height space. This can clearly be seen in the cross-section, which also shows the suspended *brise soleil* along the southern exposure.

The artificial lighting to the offices is in the form of linear fluorescent fittings recessed into each reinforced concrete coffer at right angles to the walls. In addition, the street is lit from concealed light fittings adjacent to the roof light and set below the underside of the cantilevered walkway.

Interior of central street (Photographer Dennis Gilbert).

Case study 38 British Library Reading Room, London

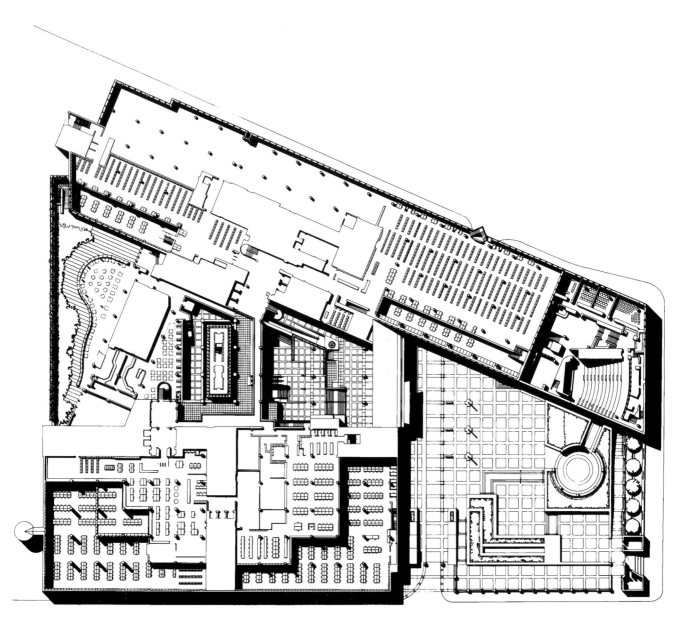

Plan of the Humanities reading room (Sir Colin St. John Wilson).

The British Library near St Pancras Station in London was designed by architect Sir Colin St. John Wilson in 1974 but was not completed until 1998. The brief for libraries in the 1970s was very different to the brief that would be given today, as can be seen by looking at the library designed for Brighton University (see Case Study 36 Aldrich Library, University of Brighton).

The earlier brief would have suggested that daylight was not desirable and that libraries called for high levels of even

Architect: Colin St. John Wilson & Partners; Lighting Design: Steensen Varming and Mulcahy

Exterior of the entrance area (Photographer Martin Charles).

Portrait view of Humanities reading room (Photographer Martin Charles).

light. Happily the architect ignored this advice; he allowed the entry and stimulation of daylight and provided variety of environment whilst satisfying the requirements for conservation of books. The need for conservation is at its highest in the area devoted to the King's Library, and this is perhaps one of the most exciting areas of the library, where the light levels are at their lowest (50 lux).

It would be impossible in this short case study to try to cover such a vast library complex so the study is limited to the main, or Humanities, reading room, which is inevitably compared to the original round reading room at the British Museum which it replaces, and where Karl Marx worked. The main difference, ignoring the advances in information technology, is that the latter offered a single all-embracing space, whilst the new library offers a choice of environment with a variety of work places and views.

Daylight is introduced through clerestory windows as the principal source of ambient light, and whilst direct sunlight is eliminated by a louvre system from the reading positions to avoid glare, sunlight is permitted to hit the sloping ceiling areas above to introduce the external environment to the space. Daylight associated with indirect up-lights provides an illumination of 250 lux without recourse to the task lights at the desks.

The ceiling surfaces are washed with indirect light from low level, and a personal desk light may be switched on to provide 500 lux when needed for reading, but which would be undesirable when a computer screen is used. The need for an extended view for relaxation of vision is catered for by the size of the space which takes the place of views out through windows.

Case study 39 Cable and Wireless College, Coventry

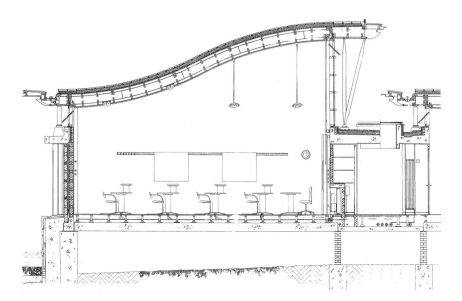

Section through one of the teaching blocks (MJP).

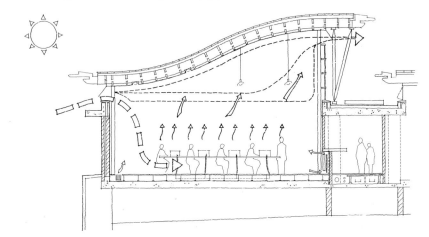

Ventilation section (MJP).

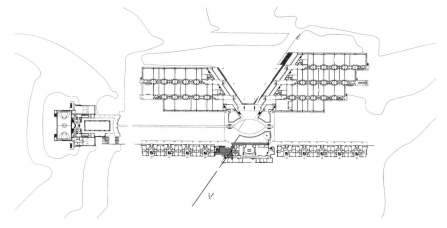

Plan of the complex (MJP).

Historically, the telecommunications college was located at the tip of Cornwall, where Transatlantic Cables first arrived. But by the 1990s a decision was made to relocate to a site near Coventry to take advantage of better transport links, and to be close to related technical departments at the University of Warwick at Coventry.

The brief to the architects MacCormac Jamieson Prichard was for a campus to train 300 students, with teaching accommodation, residential and associated recreational facilities. The client wanted the buildings to be naturally lit and ventilated.

This case study is concerned primarily with the teaching wings, which are a series of parallel buildings along the south side of the site. The class modules are 9 metres wide, with 3-metre corridors between and can be divided laterally into units of 4.5 metres to achieve varying class sizes, depending upon the teaching requirements.

A solution could have been found using traditional north light roof trusses to provide natural light into the 45-metre-deep arrangement of three classrooms but this would not have solved the problem of natural ventilation. Laboratory modelling of the wave-shaped roof section demonstrated that with cross ventilation and a stack effect the spaces could cope with the heat spill predicted.

Natural light enters the classrooms from the north and taller side, with additional light and ventilation entering from the lower and smaller windows to the south (see Section).

The artificial lighting scheme in the typical 4.5 metre module has fluorescent lamps, recessed into the ceiling on the lower side and suspended from the higher ceiling close to the taller windows. The two sets of fittings are separately controlled to allow energy savings by switching off the pendants when daylight is sufficient.

The key to the energy savings is in the means of control and the college was determined to use manual controls rather than a BEMS. The lighting controls, along with the motorized blinds, window

Architect: McCormac Jamieson Prichard (MJP);
Lighting Design: Ove Arup and Partners

End view of teaching block to illustrate section (Photographer Martin Jones).

opening gear and power controls, are related to each 4.5 metre module, allowing the partitioning to be removed and relocated to vary the room size without compromising the lighting layout.

The partitions themselves contain no wired services so they can be removed and new ones built over a weekend to provide the flexibility demanded by the teaching staff.

The whole complex is one of great elegance and sophistication and will be a benchmark for campus planning in the future. The extensive use of daylight in all the public areas provides a natural variety of experience in other spaces, including the offices, the bar, dining room and sports facilities.

Daylight view in a teaching module (Photographer Alex Ramsey).

Case study 40 Student Union Building, University of Durham

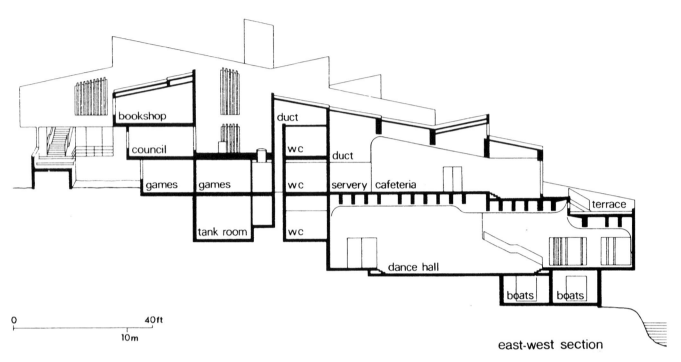

0   40ft
|————————————|
      10m

Cross-section to show the building's relationship to the slope of the gorge, and the manner of the daylight entry (ACP).

east-west section

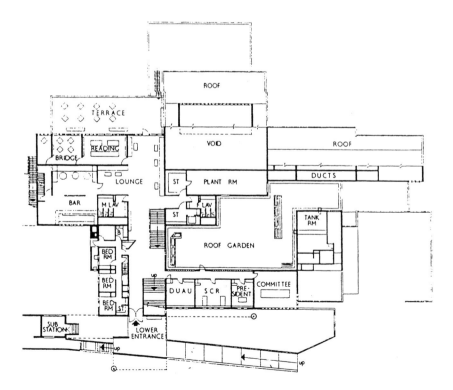

Plan (ACP)

Dunelm House, the Student Union Building for the University of Durham, was built in 1965 and designed by the Architects Co-Partnership (ACP). The building provides extensive recreational facilities including a dance hall, book shop, games room and café. It was built to accommodate a student population of 1500, expanding to 3000 by 1971, and an academic staff of 160, increasing to 300.

In addition to its magnificent setting on the banks of the river Wear overlooking Durham Cathedral, the building is approached by a bridge designed by Ove Arup that spans the river and appears to grow into and become a part of the steep gorge-like slope.

But it is not the siting, but the lighting, of this remarkable building that is the subject of this case study. The lighting displays a rare understanding of the environmental aspects of daylighting during a period when architects were beginning to feel that the future of lighting lay in electric light. Little needs to be said about the artificial lighting, other than it is integrated with the structural ceilings using recessed filament fittings.

The daylighting for the interior was designed to reduce the contrast between

Architect: Architects Co-Partnership (ACP); Lighting Design: Engineering Design Consultants

Typical view of interior space (Photographer Derek Phillips).

the internal wall surfaces and the brightness outside by using panoramic windows stretching from side wall to side wall and also installing a run of high-level windows to lighten the ceiling. The daylight gives good interior modelling.

Whilst many buildings of this period tended to shut out the view, here the wonderful view over the gorge to the cathedral is glorified; and it is possible to enjoy the changes of the season and the weather in all its infinite variety.

In the deep space of the lecture theatre, rectilinear daylight openings are recessed into the ceiling at the rear of the room to reinforce the natural light.

View from the interior of the building across to the cathedral (Photographer Derek Phillips).

View of the building from across the river showing the Arup bridge by day (Photographer Derek Phillips).

Section 8 Health

Case study 41 Wansbeck Hospital, Northumberland

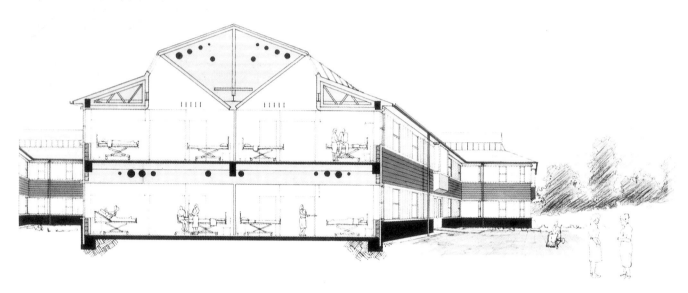

Sectional perspective through wards (Powell Moya & Partners).

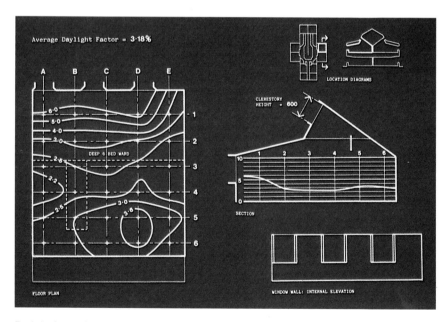

Average Daylight Factor = 3·18%

LOCATION DIAGRAMS

CLERESTORY HEIGHT = 600

DEEP 6 BED WARD

SECTION

FLOOR PLAN

WINDOW WALL: INTERNAL ELEVATION

Daylight factor diagram (Photography Powell Moya & Partners).

Wansbeck Hospital on the outskirts of Ashington in Northumberland was completed in 1993, the second National Health Service low-energy hospital following the building of ABK's St Marys Hospital on the Isle of Wight.

The brief to architects, Powell Moya Partnership, was to achieve a target of a 60 per cent reduction in the use of energy compared to a traditional nucleus plan hospital. To assist with the development and introduction of the natural lighting and ventilation systems needed to achieve this target, an increase of 8 per cent in costs above NHS limits was permitted.

There is some similarity in the cross-section through the building (see section) between the two solutions, both designed to provide good natural light into the wards at first floor level.

The roof lighting design, which uses a walk-through duct at high level, was tested under an artificial sky and provides natural light to the rear of the deep wards. Coupled with the side windows this means that electric light is rarely required even on overcast days. To eliminate glare, vertical baffles are provided to intercept the patient's view of the roof lights.

A simple system of artificial lighting using pendant fluorescent fittings can be

Architect: Powell Moya & Partners;
Lighting Design: Cundall Johnston & Partners

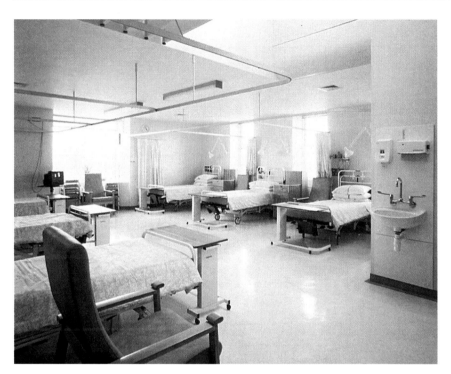

Typical ward, illustrating the natural and artificial lighting systems (Photography Powell Moya & Partners).

seen in the picture of a typical ward, with individual wall-mounted bed head fittings provided for the patients and for doctors' inspections.

The artificial light is computer-controlled by the Building Energy Management System (BEMS) to react to the level of daylight outside and the occupancy of the spaces, switching off the lights when there is sufficient natural light or when the rooms are unoccupied. Manual override is provided allowing some individual control. Individual control is also given to the ventilation system, in that during the winter the windows are locked, whilst during the summer windows may be opened by the patients.

A further innovation is the appointment of an Energy Manager, who monitors the use of energy throughout the building, keeping staff and patients aware of any overt changes in the environment. In addition a wind turbine has been installed to generate 10 per cent of the overall energy requirement.

Dining Room (Photographer John Parker, Powell Moya & Partners).

Case study 42 St Mary's Hospital, Isle of Wight

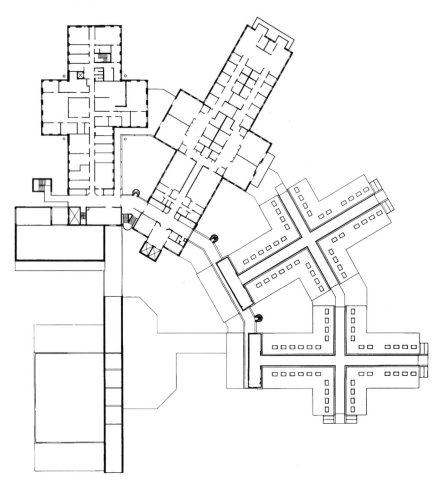

Nucleus plan of a section of the hospital (ABK).

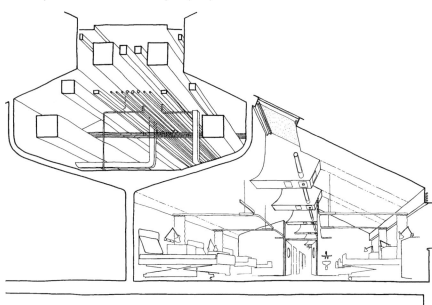

Section showing ducting and natural and artificial light (ABK).

St Mary's Hospital on the Isle of Wight, designed by architects Ahrends, Burton & Koralek (ABK), is the principal hospital for the island, offering 191 bed spaces, new outpatient's, children's and old people's departments, laboratories, workshops and operating theatres, as well as a social centre for the village.

The plan is organized on the nucleus principle, a system based on a cruciform plan, with legs containing the patients' areas and a central station for nurses. The legs can be planned in a variety of ways to accommodate small or large wards according to need. All patient areas have access to good natural light and views to the excellent landscaping of the outside courtyards.

Maximum use of natural light is a key element in the energy-saving strategy, reducing the need for artificial light. The building is artificially ventilated in the wards but opening windows help with summer ventilation, reducing artificial ventilation demand and giving patients the sense of it being in their control.

The section illustrates the service ducting above the centre of the wards, which carries all the pipes, ducts and electric cable, whilst further illustrating the way in which natural light enters both from the side windows and from roof lights following the roof contours.

After dark, artificial light is provided by a tubular fluorescent system echoing the line of the lay light above and suspended between the air conditioning outlets which divide up the space. The daylight is adequate for normal general lighting of the wards during the day, but each bed is provided with an adjustable wall-mounted lamp, used also as an inspection lamp when needed.

The patients' restaurant area has good natural light from side windows in addition to pendant glass spheres giving overall artificial light after dark.

Architects: Ahrends, Burton & Koralek (ABK); Lighting Design: Building Design Partnership

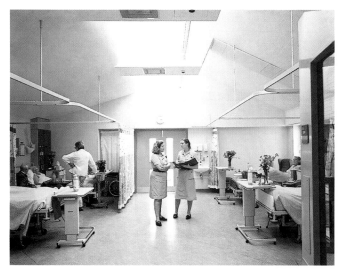

View of ward to show the natural and artificial lighting systems (Photography ABK).

View from the window to the outside court (Photography ABK).

Patients' restaurant (Photography ABK).

Case study 43 Finsbury Health Centre, London

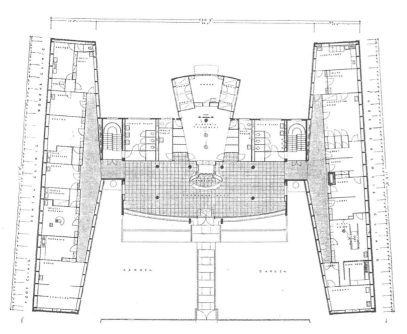

Ground floor plans (Tecton).

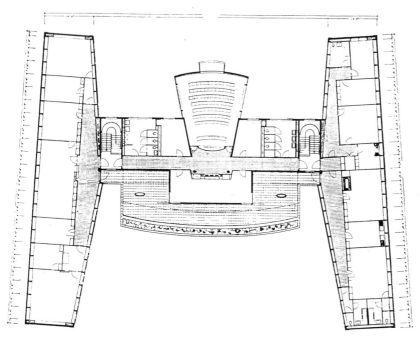

First floor plans (Tecton).

Described by Banister Fletcher as one of the 'structural and aesthetic masterpieces of English modernism,' the Finsbury Health Centre in London, designed by Lubetkin in 1938, survives to still function today in the manner it was designed. Avanti Architects have undertaken conservation and repair work in the 1990s with a view to ensuring that its purposes are well-served into the twenty-first century.

Perhaps the most memorable feature of the centre is the front foyer which, as can be seen in the plans, extends across the width of the central area of the site and is entered by a bridge across the garden. This foyer is lit by natural light during the day using a wall of glass bricks, and has changed little in the intervening 60 years, the daylighting being as successful today as when it was built.

The artificial lighting on the other hand has been updated using surface-mounted fluorescent fittings, but evidence of the original fittings survives. It can be seen in a photograph taken in 1938 that the original fittings were specially-manufactured metal shades supported on floor standards. These torchères would have provided pools of light in the sitting areas to create, in the words of Lubetkin, 'the atmosphere of a club.' Some additional light is provided by a ceiling-mounted, bullet spotlight.

The same shades as used on these torchères were also mounted on the rear wall to the foyer to provide uplight to the ceiling, and are an interesting historic reminder of the care taken by architects in the 1930s to achieve technological solutions to the new problems posed by the modern light bulb.

All the offices and consulting rooms located in the side wings are daylit and the small lecture theatre at the rear is now used as offices, where the original glass wall has been supplemented by lines of fluorescent fittings mounted at each side above the windows. The access corridors are lit during the day by side windows, giving very adequate daylight.

It is perhaps no surprise that the centre was designed to be a very energy-efficient building, where on most days the electric lighting would have only been used at night.

Original architect: Lubetkin, Tecton;
Architect 1990s: Avanti Architects

Original architect's sketches circa 1930s

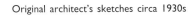

WELL LIT ROOMS

LARGE CONTINUOUS WINDOWS AND
SHALLOW ROOMS GIVE AMPLE LIGHT
IN ALL PARTS OF THE BUILDING.
BLINDS EASILY FITTED WHERE NECESSARY

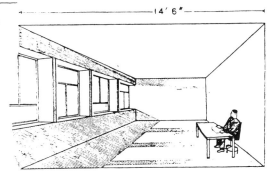

DUCTS FOR LIGHT WIRING

CONTINUOUS DUCTS WITH
REMOVABLE COVERPLATES
IN THE CEILING GIVE POS-
SIBILITY OF ANY NUMBER
OF ADDITIONAL FITTINGS.
EASY ACCESS FOR REPAIRS

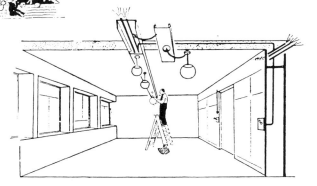

Original sketches of the artificial lighting (Tecton).

Perspective of open planning (Tecton).

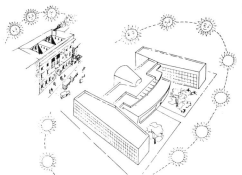

The foyer, 1938.

The foyer, 1999 (Photographer Derek Phillips).

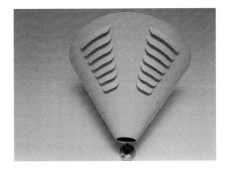

The torchère shade used as an uplighter in the foyer (Photographer Derek Phillips).

A typical daylit corridor (Photographer Derek Phillips).

Section 9 Shops/display

Case study 44 Treasures of Saint Cuthbert, Durham Cathedral

TREASURES OF SAINT CUTHBERT

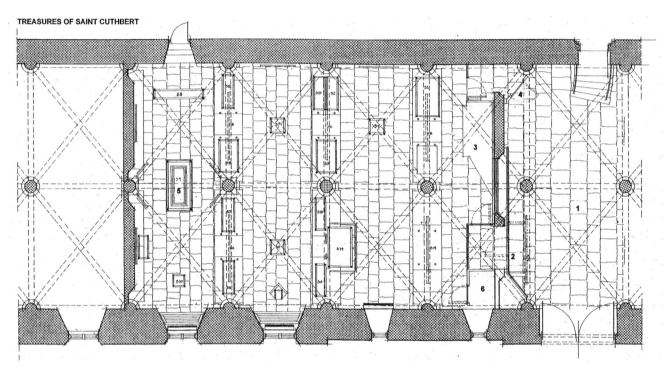

Plan of the undercroft (Studio E).

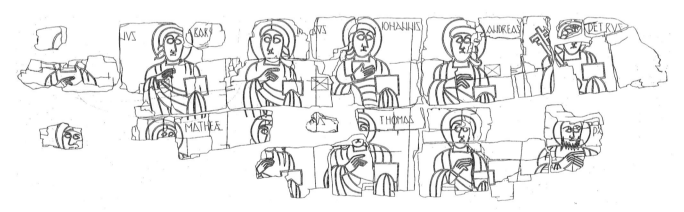

Seventh century coffin carving (Studio E).

The exhibition of the Treasures of Saint Cuthbert at Durham Cathedral is seen not so much as an art collection (although of course it is), but as an example of careful display to show off the artefacts to their advantage in their unparalleled location in the vaulted undercroft of the cathedral cloisters.

The collection consists of Saint Cuthbert's coffin, his pectoral cross, the seventh century Durham Gospels and the Anglo Saxon embroideries of 935, one of several gifts from King Athelstan. The

Architect: Studio E Architects;
Lighting Consultant: Sutton-Vane Associates

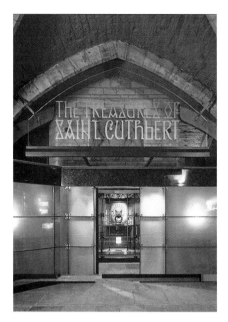

Exhibition entrance (Photography Studio E).

Detail to show the knocker (Photography Studio E).

Coffin of Saint Cuthbert (Photography Studio E).

display therefore had to have a logical sequence and at the same time satisfy the needs of conservation.

The work, finished in 1998, took two years to complete, in the hands of a design team led by the Studio E. Architects, which included historians, archivists, the curatorial staff, graphic designers, craftsmen, lighting consultants and the Professor of Archaeology from Durham University.

The main problem in lighting a display of this significance is to maintain the low light levels required for conservation whilst at the same time creating the impression for the visitor that the artefacts are not dimly lit. One way to achieve this is by adaptation, starting with artefacts such as gold and silver metals which can be highly illuminated and gradually reducing the levels towards those items such as embroideries, mediaeval fabrics and manuscripts, where the lowest light levels are mandatory.

Much use is made of edge-lit glass for signs and historical information, and at the front entrance back-lit panels create a bright impression against a darker background to meet the design brief for an attractive entrance. The panels change colour, running through a complete dawn to dusk sequence.

For the displays themselves, each has a lighting system designed to suit its specific requirements. Graphics panels have spotlights to their top, whilst well-shielded, low-voltage spots are used to highlight each artefact individually. The spots had special filters designed to eliminate the adverse effects of ultraviolet light.

The lighting of the gothic columns and vaulting of the undercroft was achieved using fibre optics set into the floor, lighting upwards to bring out the texture of the stonework and to enliven the vaulting above.

The sequential experience of the display is designed to reinforce the notion of travelling back through time starting with the brilliant seventeenth century cathedral plate and returning by stages to the earliest stages of the Christian church with the centrepiece of the coffin of Saint Cuthbert.

Case study 45 Harlequin Shopping Centre, Watford

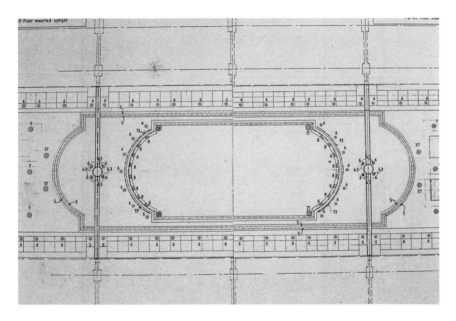

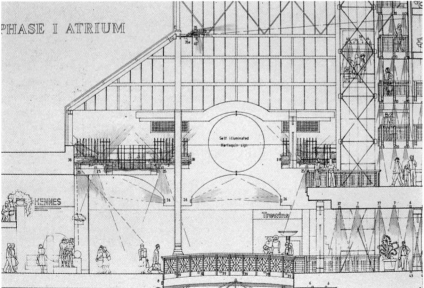

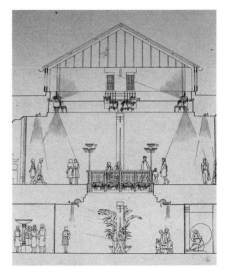

Typical plans and sections, showing the lighting design (DPA).

Built at the beginning of the 1990s, the Harlequin Centre in Watford, designed by architects Chapman Taylor Partners, represents a combined daylight and artificial lighting strategy for an urban shopping centre, built at a time when it was more fashionable for large centres of this kind to be located on the outside of towns.

The site itself was restricted by existing street patterns, so that the orientation of the centre had to conform, whilst at the same time the centre was required to connect with an existing shopping complex on an adjacent site.

The daylight strategy takes advantage of the opportunity to use overhead roof lights which are planned to infiltrate natural light deep into the interior of the building. The complex relationship of levels and voids, which can be seen in the accompanying sections, ensures that daylight is available in all the public areas.

There is little doubt that the available daylight creates spaces which are pleasant, and which, because of the size of the project, permit internal views which make up for the lack of views out to landscape at a distance. The spaces are arranged in such a way that a visitor is

Architect: Chapman Taylor Partners;
Lighting Design: DPA Lighting Consultants

Daylit interior of the mall (Photography Derek Phillips).

Night view of the light troughs (Photography Derek Phillips).

Night view of the mall (Photography Derek Phillips).

aware of his orientation, where he is and the state of the sun and sky and can enjoy natural colour and the associated variation of environment due to the weather.

The general artificial lighting at night employs low-wattage, warm and cool metal halide lamps in fittings recessed into the suspended ceilings. These are supplemented by decorative pole-mounted lamps located around the edge of voids, rather similar to the small-scale street lamps used in urban pedestrian areas. The recessed fittings are located in groups to provide areas of focus and avoid the more arid impression created by an overall level of light. The underside of certain voids were developed as concealed light troughs, to soften the contrast between natural and artificial light.

The tenants of the shops themselves designed the artificial lighting which best suited their merchandise, leading to an infinite variety of solutions. It is this variety, duplicating the original town high street that makes the experience of this type of shopping street most acceptable.

The artificial lighting is controlled by a Building Energy Management System (BEMS) designed to reduce the energy use. The BEMS relates the level of artificial lighting to that required by the rise and fall of daylight, ensuring that when the building is unoccupied sufficient light is employed for security, whilst cleaning lights are used only at levels and in areas for functional needs. An important element of the overall strategy is that there is a distinct variation between the day and night ambience of the centre, conforming to our normal experience of day and night.

Case study 46 Erco showroom, London

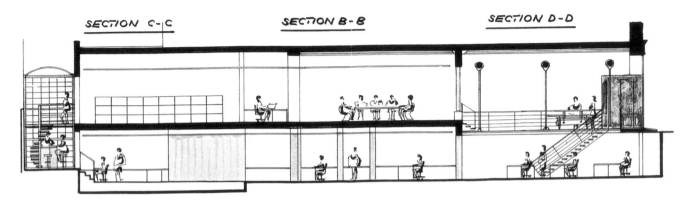

Section through showroom (Erco Lighting).

View across bridge to frontage at night (Photography Erco Lighting).

The London showroom for the lighting manufacturer, Erco Lighting Ltd, is both an example of display lighting and the display of lighting and it is successful on both counts.

The architect Pierre Botschi recognized that daylight is an essential ingredient of design, even when as here it is available only at the front and rear of a deep-plan London property.

On entering the showroom one is confronted by a light void spanned by an illuminated glass bridge leading into the showroom beyond. The void allows daylight to penetrate down to the lower office level, where light is reinforced where required by task lighting.

The main showroom in the centre of the space is lit artificially by fittings manufactured by the company. These act both as a flexible lighting installation for a variety of displays and as display of the equipment itself. At the rear of the property daylight is again welcomed as each level of the building is connected to a garden terrace beyond.

On entering the showroom the impression is one of well daylit space where the display of equipment is not overwhelming, but which at the same time is sufficient to demonstrate adequately the manufacturer's products.

Architect: Pierre Botschi; Lighting Design: Erco Lighting

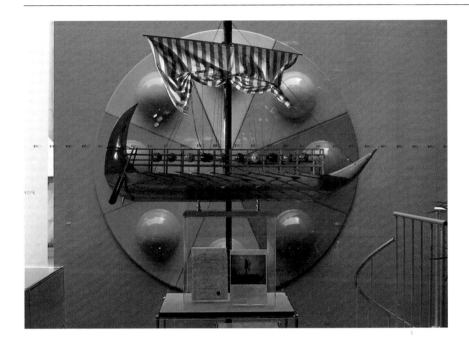

View to Verner Panton's roundel with ship display at the rear of the showroom (Photography Derek Phillips).

View of the display of 'The Victory' during a special exhibition of ships in the showroom in 1998 (Photography Derek Phillips).

Case study 47 John Lewis store, Kingston-upon-Thames

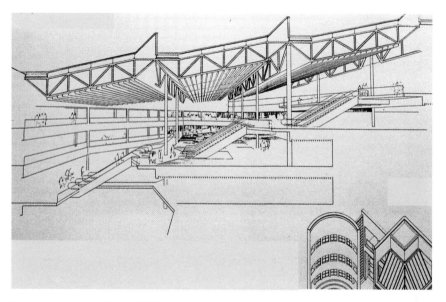

Section to show daylighting (ABK)

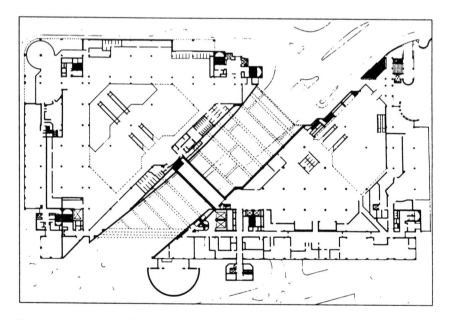

Plan to show road way below building (*Electrical Design*, June 1993).

The John Lewis store is sited on the banks of the Thames at Kingston-upon-Thames, Surrey. Designed by Ahrends Burton & Koralek in the 1980s, the building was completed in 1993.

The lighting strategy for the building is one of daylight during the day supplemented where necessary by artificial sources. The goods displayed are enhanced by the colour and variation of daylight which reaches down through five storeys from a 3200 square metre glazed roof structure.

The roof consists of 'solaglass' double-glazed units incorporating a capillary diffusing layer of Okalux which reduces the solar gain and ultraviolet transmission. A small section of clear roof glazing at an angle of 45 degrees allows the impression of sunlight to enter; control is by electrically-operated helioscreen blinds.

The daylight is designed to provide a minimum of 500 lux, and when the level falls below this the computer-controlled artificial lighting cuts in. The artificial sources are 250 watt HQI lamps in specially-designed reflectors.

The shop has all the attributes of a daylit space, but some areas do not receive daylight. In side areas special transition zones have been planned in the perimeter walkways. Here a reflective floor surface and a high level of artificial light help the visual transition from what can be as much as 4000 lux of daylight on the sales terraces to 500 lux below the lower ceilings.

In an age where the tradition in large stores had been for totally artificially-lit interiors the John Lewis Store was ahead of its time in satisfying the needs of the environment in an energy conscious fashion.

To study the daylight and sunlight penetration into the building the Austrian consultant Friedrich Wagner constructed a large-scale daylight model of a section of the roof in Vienna; thus proving the effectiveness of model studies over calculation techniques in some cases.

Architect: Ahrends, Burton & Koralek (ABK); Lighting Consultant: Friedrich Wagner, Vienna

Exterior view from river (Photography Derek Phillips).

Daylight view of the interior (Photography Arup Associates).

Night-time view (Photography Derek Phillips).

Section 10 Art galleries

Case study 48 Daylight Museum, Japan

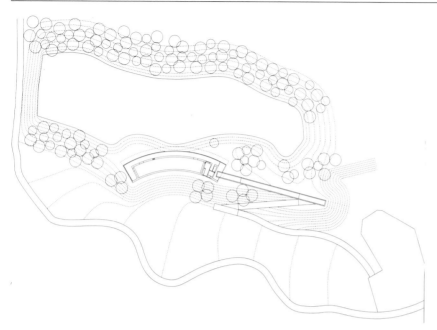

Plan of art gallery together with the lake (Tadao Ando).

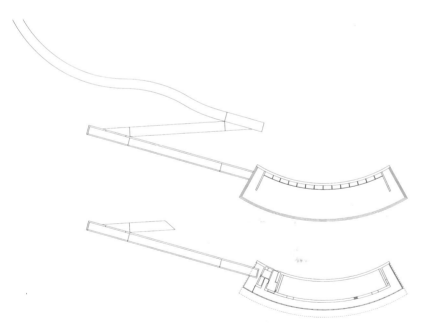

Plan of the roof light (Tadao Ando).

The art gallery at Gamo-Gun Shiga in Japan designed by architect Tadao Ando is lit by daylight only and closes once daylight fades at sunset. The architect remembered an artist working after the last war in conditions of natural light with all its variety, and ceasing to paint once the light had gone. So it was that when asked to build a gallery for the work of this artist many years later, he decided that only natural light would do.

The gallery is built at the side of a lake, with a curved façade that echoes the shape of the lake. The façade consists of a frosted glass curtain wall that delivers changing light to the corridor beyond.

The main gallery located behind the corridor is lit entirely by top lighting from a circular arc in the roof. The quantity and quality of natural light entering the gallery changes with the time of day and the seasons, offering diverse impressions of the exhibits where visitors may obtain quite a different experience on different trips to the gallery. This is very evident from the views illustrated.

It has to be said that this approach goes against the trend of modern art galleries with their preoccupation with preserving the works of art for posterity. Such ideas of conservation have no place here, the architect taking the view that 'paintings are not to be suspended in time, sealed and worshipped once the artist applies the final stroke.' This approach would have been commonplace in the seventeenth century, but it should be borne in mind that many paintings of the early periods have suffered in the past from exposure to sun and day light, and there is a balance to be struck.

Architect: Tadao Ando, Japan

Exterior with lake (Photographer, Kaori Ichikawa, Tadao Ando).

Interior of gallery suffused with sunlight (Photographer, Kaori Ichikawa, Tadao Ando).

Interior of gallery showing the daylight slot (Photographer, Kaori Ichikawa, Tadao Ando).

Case study 49 The Burrell Collection, Glasgow

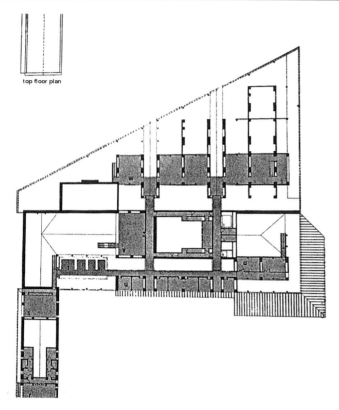

top floor plan

Overall plan of building (provided by the Burrell Collection).

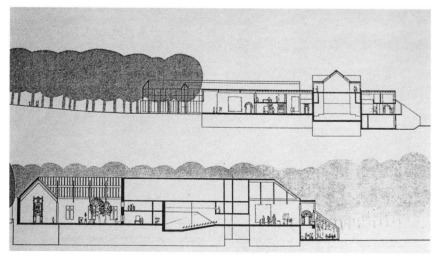

Section through building (provided by the Burrell Collection).

The Burrell Art Gallery was the result of an architectural competition won by architect Barry Gasson to design a gallery to house the works of art collected by Sir William Burrell and given on his death to the city of Glasgow. Being a finite collection, the gallery was designed around known works of art, and for this reason it did not have to provide for the sort of flexibility normally associated with public art museums.

The building is a celebration of daylighting, which enters the building through wall-to-ceiling glazing along the north side. The glass wall is protected by a belt of trees that acts as a screen to ensure that only limited daylight enters the museum, the trees themselves catching the sun when available. Daylight enters through roof lighting in the sculpture court and the cross corridors, and is also used to illuminate a collection of stained glass displayed along an exterior corridor.

The architect has used daylight skilfully wherever the exhibits allow. Where strictly limited levels of lighting are needed for the exhibition of tapestries and other biodegradable objects the exhibits are housed in areas away from the external walls and lit entirely by artificial light at a low level of 50 lux.

The artificial lighting is derived from spotlights located in the timber roof construction. There is no attempt to recess or conceal the lamps as the structure itself is of a robust nature and can accept this form of lighting design. The fittings can be directed and controlled to provide the best lighting for the exhibits on display.

The result of the architect's design is to provide a sequence of experiences related to the site, in which daylight appears to be the main light source during the day, whilst artificial light sources are used to ensure the right balance of light for the exhibits, maintaining strict conservation standards in those areas where sensitive materials are shown.

Architect: Barry Gasson

Sculpture court (Photographer Derek Phillips).

Interior to show north wall (Photographer Derek Phillips).

Cross corridor with overhead light (Photographer Derek Phillips).

View through north wall to sunlit trees (Photographer Derek Phillips).

Artificially-lit gallery for tapestry (Photographer Derek Phillips).

Case study 50 Carré d'Art, Nimes

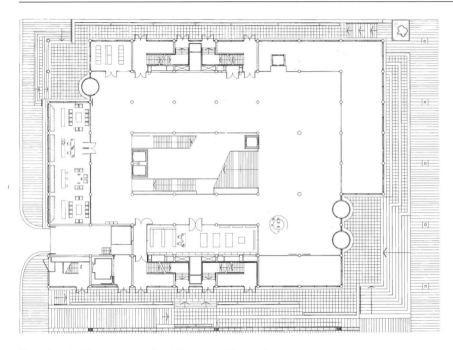

Plan of the building at ground level (Foster and Partners).

The Carré d'Art at Nimes, designed by Sir Norman Foster and completed in 1993, is a complex containing a museum of contemporary art and public libraries together with numerous ancillary facilities such as cinema, conference centre and storage areas.

The brief to the architect was to design a building which would contrast with but not compete with the neighbouring First Century Roman Temple, the Maison Carrée, where each would enhance the other, to refer but not to defer.

In addition to designing the building itself, the architect planned to re-route traffic to create a large pedestrian area around the museum as a part of the general pedestrianization of the town.

While fulfilling a variety of functions, the building maintains roughly the same overall height throughout by sinking four of its nine storeys below ground, with a central daylit atrium containing glass staircases which encourages daylight to the lower levels.

The natural light from the roof level is filtered by mechanized louvres above the glass with adjustable diffusing vanes below. Fixed sun-control louvres are used throughout the building, beyond conventional thermal-pane glass, much of which is silk-screened white to shade the galleries and services. The gallery areas are diffused with daylight, some of which is modified by entry through the sides of the central atrium to reduce the level of light for reasons of conservation.

The artificial lighting is by lines of ceiling mounted fluorescent lamps, with track mounted filament spots picking out salient features.

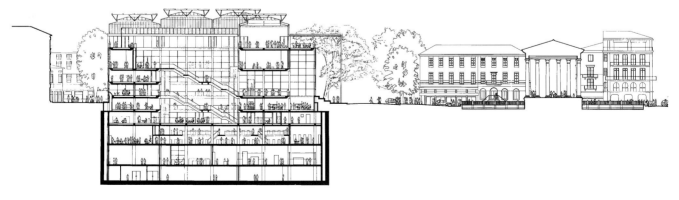

Longitudinal section (Foster and Partners).

Architect: Foster and Partners; Lighting Design: Claude Engle

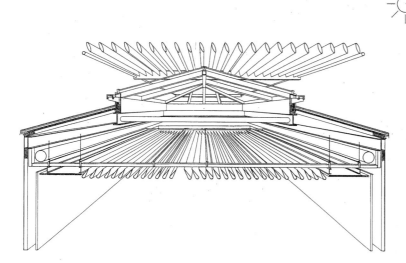

Sunlight-control louvres to gallery skylights (Foster and Partners).

Central staircase atrium (Photographer James H. Morris, Foster and Partners).

Daylit gallery to internal courtyard (Photographer Dennis Gilbert).

Library interior (Photographer James H. Morris, Foster and Partners).

Case study 51 National Gallery, Sainsbury Wing, London

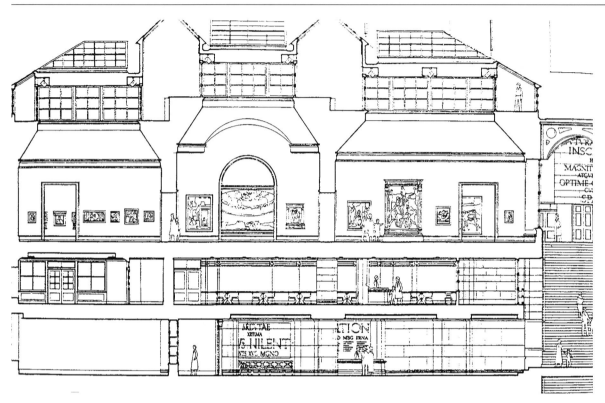

Transverse section (VSBA)

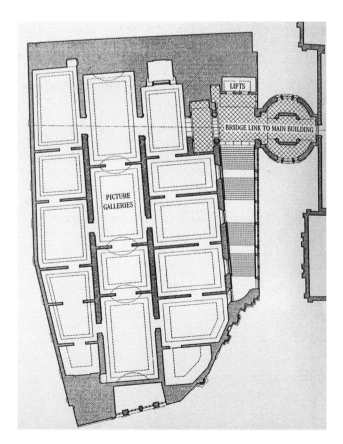

Plan (VSBA).

An extension by architects Venturi, Scott Brown and Associates to the original National Gallery, designed by William Wilkins in 1838, was completed in 1991. The building stands on its own in Trafalgar Square, but is connected by a bridge to the main series of galleries.

The plan shows the way in which the space is broken up into a series of small galleries, to provide the necessary wall space for the paintings. Although there are three main levels and a basement, the galleries themselves are all planned at the top level, to take advantage of the natural light; indeed it is only when looking upwards from Trafalgar Square towards the roof level that one is aware of the building's daylight credentials, since the general appearance is of blank walls. Some windows give side lighting at the lower levels to the main entrance hall, grand staircase and ancillary accommodation.

It is the nature of the lighting of the galleries which is of importance in the context of this case study, for whilst the galleries appear to conform to traditional overhead daylighting, the solution is one in which the daylight contributes to the environmental appearance of the galleries,

Architect: Venturi Scott Brown & Associates; Lighting Design: Fisher Marantz Inc, USA

General view looking along the length from one gallery to another (Photography VSBA).

Vista to show Renaissance painting (Photographer Matt Wargo, VSBA).

View at roof level showing daylight detail, with access to catwalks (Photographer David Loe).

emphasizing its natural variation and recognizing its changeability.

To ensure that the levels of daylight on the paintings are consistent with acceptable standards for conservation, a 'hi-tech' solution has been adopted where unlimited daylight is permitted into the top of the building, but is then computer-controlled by a system of louvres according to the level of daylight outside. It can be seen in the picture of the detail at roof level that no direct daylight or sunlight reaches the paintings, but the impression of the daylight and sunlight outside is still experienced by the visitors to the galleries.

The artificial lighting system consists of low-voltage tungsten halogen fittings placed at high level on track around the roof lights and controlled to supplement the daylight and to provide a warmth to its colour. It is the combination of the daylight and the artificial light which is used to light the paintings to an average level of 200 lux. This level is allowed to vary slightly provided that the total does not exceed 650 000 kilolux hours in a year, as logged by computer.

When the galleries are closed, from 6pm until 10am, the daylight louvres are closed and the pictures left in darkness, apart from a small level of security light this further reduces the total kilolux hours of light.

The rooms are designed to exhibit Renaissance paintings, rich in colour and texture, and the combination of cool, natural light with warmer, artificial sources, is most successful. A visitor is aware of the weather and time outside with some views out from the grand staircase, the three identifiable states – bright daylight, dull days and night time – providing a desirable degree of variety.

Case study 52 Tate Gallery, St Ives

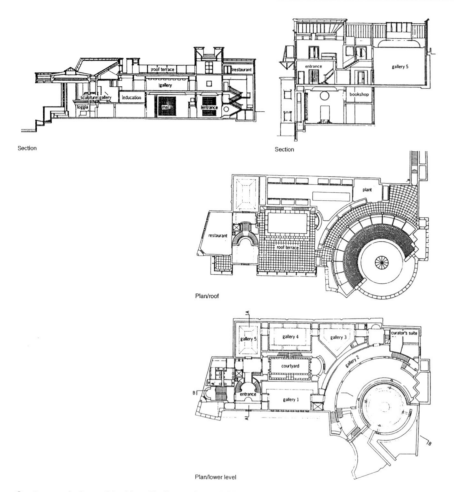

Section

Section

Plan/roof

Plan/lower level

Sections and plans of building (*Architects Journal*, 23 June 1993).

Exterior of building showing its close relationship to the town and harbour (Photographer Derek Phillips).

The Tate Gallery at St Ives in Cornwall was completed in 1993 and was built first and foremost to house the works of art of the St Ives school of artists who worked in the area in the period between the two world wars.

The building was designed by architects Evans and Shalev to be an integral part of the fabric of St Ives with its views out over the harbour, a building that was to be experienced in the natural surroundings in which the artists worked. The site was a special challenge as it was built on a hill side falling 12 metres down the cliff face.

As with all modern art galleries, the importance of providing space for the enjoyment of works of art under controlled environmental conditions meant that the introduction of daylight required careful consideration, if deterioration of the works was to be avoided.

A series of four galleries planned around a central courtyard are in the words of the architects 'spaces with varying shapes, sizes and quality of light, to highlight the paintings and objects and create a relaxed atmosphere conducive to experiencing the art.'

Daylight is derived from overhead roof lights incorporating ultraviolet filters and equipped with blinds for daylight and sun screening. The levels of light planned were 150 lux for walls displaying oil paintings, going down to 50 lux for watercolours.

The most dramatic space is the semi-circular gallery for the display of sculpture. Planned on two levels, the admission of daylight ensures a variety of experience depending upon the external conditions, but sunlight is controlled for reasons of conservation and visual comfort.

One of the aims of the gallery is to introduce children to the joy of painting and a special work/demonstration room has been provided, whilst on fine days children can use the roof level for painting competitions.

The whole building succeeds at many levels: it is related to the town and the sea with the views from its restaurant and sculpture gallery; it relates to the people in providing the town with a significant gallery for the display of the work of local artists; and it succeeds at providing an environment for the safe display of its works of art. It is a joy to visit.

Architects: Evans and Shalev; Lighting Design: Max Fordham & Partners

The semi-circular sculpture gallery, with views out over the sea (Photographer Derek Phillips).

Patrick Heron glass wall in entrance (Photographer Derek Phillips).

Case study 53 The Grand Louvre, Paris

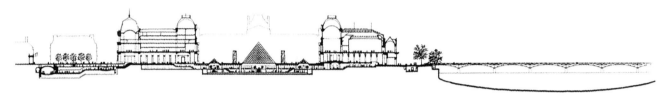

Section through building (I. M. Pei).

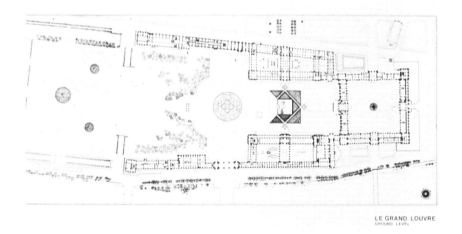

LE GRAND LOUVRE
GROUND LEVEL

Plan of the square in the Grand Louvre (I. M. Pei).

The glass pyramid at the Louvre designed by I. M. Pei of the architects Pei Cobb Freed & Partners is perhaps the most dramatic example of Gio Ponti's 'second aspect' of architecture, its appearance after dark related to its daytime appearance.

The 21-metre-high glass pyramid is the only evidence above ground of a huge development at the Louvre Museum, instigated on the orders of President Mitterand: it is a total reorganization of the 'U'-shaped building around a focal courtyard with the pyramid at its centre.

Now that the pyramid is there, it is difficult to imagine any other structure that would fit so organically into the square, as though it had been there all the time. It serves as the new main entrance to the museum, providing direct access to the three wings of galleries, in addition to providing natural light to 47 000 square metres of space below.

The space below ground contains the central reception, shops, restaurant and other support facilities, and it is the pyramid that acts as a giant roof light that brings the whole concept together, uniting it under the sky and varying its effect as the weather and the light outside changes until it fades altogether and the artificial light takes over – a wonderful sequential experience, and one that changes on each visit.

Architect: I. M. Pei of Pei Cobb Freed & Partners; Lighting Design: Claude Engle

Night view of the square with the lit pyramid (Photographer I. M. Pei).

The artificial lighting strategy was to light up only the stainless steel framework supporting the glass above. This was accomplished by developing special narrow-beam fittings to accept the 100 watt 12 volt tungsten halogen lamp. These lamps are concealed in a duct around the inner base of the pyramid and light upwards to illuminate the steel work from top to bottom, without causing interreflections in the glass panels.

The lighting of the large foyer below ground is by a variety of fittings recessed into the concrete coffers that surround the central area, but as the functions in these areas are varied, a variety of different fittings was designed to fit into the same openings to give the necessary light distributions. These were designed to provide wall washing, accent lights and adjustable downlights. The objective of the artificial lighting was to minimize the difference between the natural light in the centre and the surrounding area where less daylight penetrates.

View through structure to the buildings in the square (Photographer Derek Phillips).

View of the Louvre entrance area from the staircase during the day (Photographer Derek Phillips).

Case study 54 Kettles Yard, Cambridge

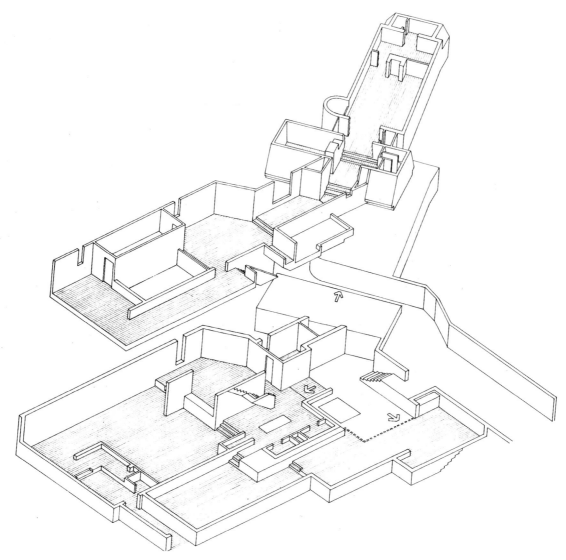

Isometric (David Owers)

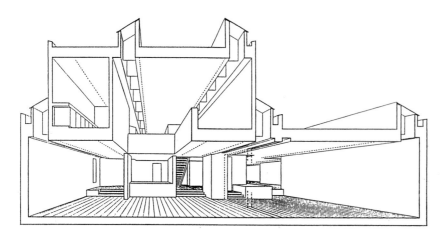

Sectional perspective (David Owers).

Good solutions to lighting an art gallery are not only associated with the great national collections: Kettles Yard in Cambridge as extended by Sir Leslie Martin and David Owers in 1970 serves as an example to demonstrate that size and standing are not necessarily the only factors.

Jim Ede, who created the original display of works of art in his own home, built the extension in sympathy with his own house conversion and it carries forward certain ideas about the manner in which works of art may be integrated within a home. The intention is to incorporate the perceptual ideas whilst enabling the enlarged collection to be viewed through a sequence of descending levels and increasing volume. For this to

Architect and Lighting Design: Sir Leslie Martin and David Owers

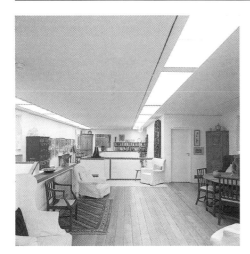

General view of interior from entrance (Photographer Derek Phillips).

be achieved, carefully scaled space and quality of light were essential ingredients.

In a home atmosphere it would be expected that light would enter from the side wall by means of windows, but this would have limited the wall space available for works of art, and an early decision was made to allow natural light in from roof level. To achieve this, long apertures were formed in the ceiling, running the length of the building. These provided the major source of natural light and were designed to light not only the top floor, but through a void, the floor below, which is evident in the sectional diagram.

The daylight design is of a permanent nature whilst the artificial light source, concealed in the same roof slots, is more ephemeral, allowing changes to be made to the mood required, and making it possible to exploit new light sources as they are developed. By placing the artificial lighting in the same slots as the natural light the overall ceiling appearance is similar both during the day and at night, providing the least obtrusive transition at dusk or when the sky becomes overcast.

General view of interior looking towards entrance (Photographer Derek Phillips).

View of the lower level, with void above (Photographer Derek Phillips).

(Photographer Richard Einzig).

Case study 55 Christ Church Picture Gallery, Oxford

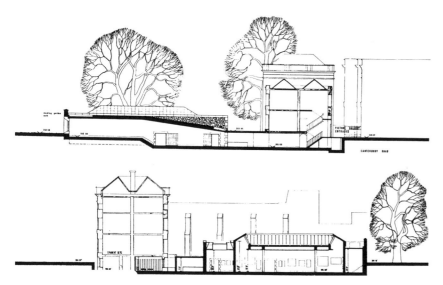

Section through the main gallery and entrance corridor (Powell & Moya).

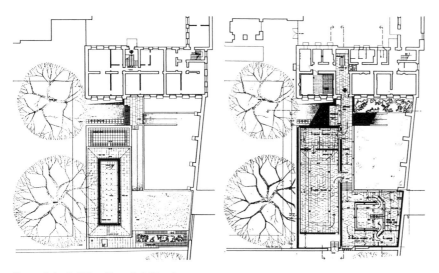

Plans of the building (Powell & Moya).

Exterior from the Dean's garden (Photographer John Donat).

Designed in 1966 by Powell & Moya, this little art gallery was required to house the painting and drawing collection of the University of Oxford's Christ Church College. Space was found on a very limited site at the end of the Dean's private garden.

The gallery has been designed with a restricted height, achieved by sinking the building 1.2 metres below ground level to ensure that the gallery is not visually intrusive. Whilst the entrance corridor has a long window overlooking the garden providing welcome daylight to the space, the galleries themselves are all top lit by daylight roof lights designed to exclude sunlight. The building is air conditioned.

The light sources, both natural and artificial, arranged so that the viewer is not consciously aware of them, are designed so that the light strikes the pictures at a high angle to minimize reflections. The daylight is controlled by a series of louvres, to ameliorate the problem associated with too much light at the top of the display wall. The artificial light uses colour matching fluorescent lamps set within the louvres together with a limited number of filament reflector spots to add warmth and accent to the paintings; these lamps are set within the daylight openings and directed at the height of the paintings, ensuring that the centre of attention is focused on the works of art.

The accommodation is divided between one large gallery, which has daylighting around the edges of a central lowered area and several smaller galleries, also top lit by daylight. A lower room has been provided for the display of prints from which daylight is excluded, the artificial light being controlled by timer to ensure that the lights are on only when the room is in use.

The general daylight strategy for the galleries has been successful, taking account of the work of Gary Thompson in limiting the levels of light on the paintings to offset the deleterious effect of high levels of daylight. However perhaps the most successful aspect of the building is the way in which the architects have integrated a substantial art gallery into the fabric of Christ Church College in Oxford.

Architect: Powell & Moya; Electrical and Lighting: Peter Jay & Partners

Corridor entrance showing view out to the Dean's garden (Photographer John Donat).

Smaller gallery showing Italian primitives (Photographer John Donat).

The main gallery (Photographer John Donat).

Case study 56 Calouste Gulbenkian Museum, Lisbon

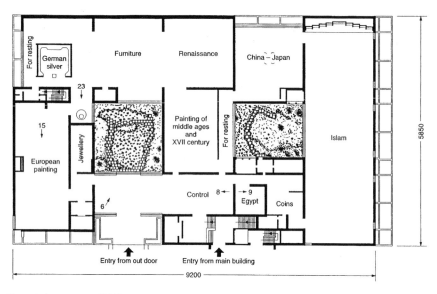

Plan of the museum (BAP).

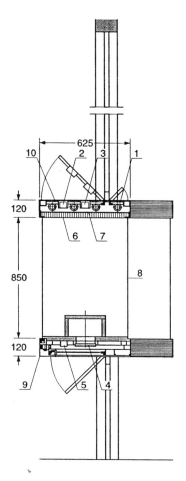

Detailed section through one of the display cases (BAP).

The design for the Calouste Gulbenkian Museum was won in competition by the architects Cid Pessoa and d'Athouguia and the building opened in Lisbon in 1969. The museum was built to house the private art collection of the Turkish collector whose name it bears, and which, apart from the art collection, includes a concert hall, conference facilities and library together with the office headquarters of the Calouste Gulbenkian Foundation.

The collection is very varied, containing some 1000 pieces, beginning in Ancient Egypt and going up to the Impressionists, and includes paintings, sculpture, furniture, manuscripts, books, tapestries, carpets, wall fabrics, ceramics, glass, silverware, coins, ivories and jewellery. Perhaps the most important aspect of the display is that it is a finite collection, and one in which each item may be shown in its best light. Flexibility, one of the difficulties experienced with temporary collections, is not an issue, allowing each exhibit to be tuned to achieve its maximum impact.

The initial idea for the lighting of the museum was to provide overhead daylighting, but before such a strategy was implemented in 1961 a work by Gary Thompson, of the National Gallery, on the conservation of organic materials was published showing that the illumination levels which would have resulted from such a method would have exceeded recommended limits. For this reason a decision was made to allow daylighting into the building from side windows, but which could be controlled to allow more daylight into those areas where conservation was not an issue and eliminate daylight where it was.

An important issue was that of adaptation since the visual impression of the lower lighting level of 50 lux required for drawings and textiles would have appeared unacceptably low if entered from a well-daylit area. Great care was taken in the planning of the spaces, to develop a sequential experience allowing the eye to adapt naturally to the desired illumination levels for each type of display. A good example of this is in the

Architects: Cid Pessoa & d'Athouguia; Lighting Design 1960: Bill Allen of Bickerdike Allen Partners; 1998: Bill Allen in association with Paul Ruffles of Lighting Design and Technology

Diana statue, daylit (Photography BAP).

lighting of the statue of Diana, where a high level of daylight is allowed from a side window.

The artificial lighting was provided in two ways: by fluorescent fittings placed between the ceiling panels or behind timber louvres; or by filament profile spots. Fluorescent lamps alone were used for paintings and ceramics, with filament being reserved for fabrics, books and parchments.

A unique feature of the design process was the use of a model of the layout where the exhibits were created in miniature so that the lighting effects could be modelled and the lighting altered to ensure the correct balance and effect.

After 30 years the same consultant responsible for the original design was commissioned to upgrade the artificial lighting in line with development in modern light sources. A room-by-room study was made of the existing lighting to judge how this might be improved without changing the overall feeling of the spaces.

In the painting galleries new tungsten-halogen spots were added to highlight each painting, and most existing tungsten lights were changed to tungsten-halogen. In all areas where fluorescent lamps were used above the louvred timber ceilings or in show cases they were changed to triphosphor lamps with modern control systems. Some adjustable fibre optics were employed to simplify the lighting in the showcases and to provide greater emphasis on the exhibits. The finite nature of the gallery exhibits was again a great advantage as each exhibit could be dealt with individually.

Islam room, general lighting (Photography BAP).

Islam room, re-lit in 1998 (Photographer Jeremy Cockayne).

Section 11 Institutions/public buildings

Case study 57 Bexhill Pavilion, Sussex

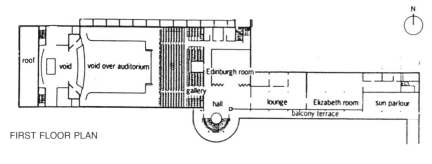

FIRST FLOOR PLAN

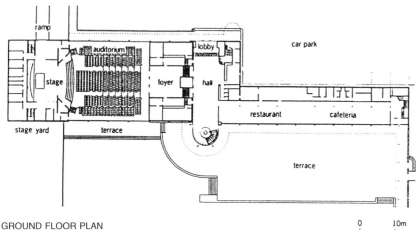

GROUND FLOOR PLAN

0   10m

Competition plans, 1934 (Eric Mendelsohn and Serge Chermayeff).

Exterior showing the central staircase (Photographer Derek Phillips).

Built by Eric Mendelsohn with Serge Chermayeff in 1934, the De La Warr Bexhill Pavilion was a monument to the Modern Movement in Britain.

After over 60 years the building had fallen into some decay caused by weathering and salt water and the architect John McAslan was appointed to develop a plan for the continued use of the building, to revitalize it architecturally and socially.

Whilst development work has still not been completed, the building is very successful and is used extensively. The principle element of accommodation is the auditorium, which has seen little change, but other elements include a restaurant serving 70 000 meals a year, a club room, and an area which in the afternoons serves as a dance hall. Other areas include a space for art exhibitions and a library.

Architecturally, perhaps the most memorable feature is the central staircase, lit during the day by natural light entering from a large, surrounding, circular window, whilst at night a 7-metre-long pendant light fitting drops through the centre down from high level. Originally this special fitting used the early 1 metre tubular lamps, but these have now been replaced with modern, slimmer fluorescent lamps. The general artificial installation consists of Bauhaus-style spherical fittings mounted at ceiling level, and these have been retained as a matter of history.

The Bexhill Pavilion was a landmark building at a time when daylight was considered as the functional light for any building. It has remained the case with natural light dominating all areas and the original artificial lighting remaining virtually unchanged, except in the auditorium where more modern fittings have been inserted into the original coffers.

Architects (original 1934): Eric Mendelsohn and Serge Chermayeff; Architects for the repairs 1991–1997: John McAslan & Partners

Interior view of the staircase at first floor level (Photographer Derek Phillips).

Detail of staircase pendant from below (Photographer Derek Phillips).

Case study 58 Ismaili Centre, Kensington, London

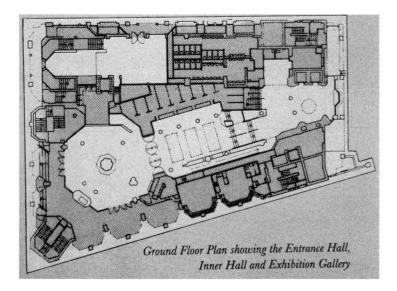

Ground Floor Plan showing the Entrance Hall,
Inner Hall and Exhibition Gallery

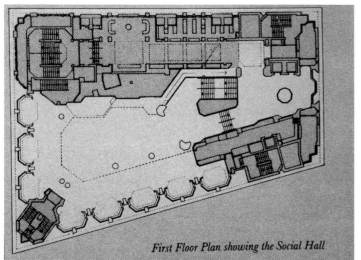

First Floor Plan showing the Social Hall

Second Floor Plan showing the main Prayer Hall

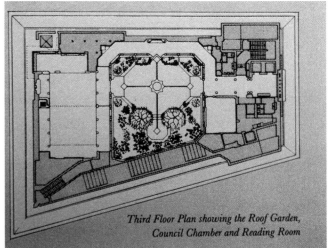

Third Floor Plan showing the Roof Garden,
Council Chamber and Reading Room

Essentially a public building for the Ismaili Muslim community built by the Aga Khan Foundation and designed by architects Casson Conder Partnership, the Ismaili Centre in Kensington demonstrates clearly by its interior design a daylighting strategy, and by its exterior the nature of the accommodation within.

The windows at the first floor level provide daylight to the main accommodation level, the bevelled-edged panes adding sparkle to the interior. At night the same windows provide an interest to the view from outside.

In contrast to this, the next level, occupied by the prayer hall, has minimal daylight received through slit windows, which give a characteristic appearance from outside and yet are obscured from the inside by Islamic timber screens.

The council chamber and library are planned at roof level together with the roof garden. Special roof openings allow in the daylight, whilst artificial sources are concealed in the same location.

The building displays an integration of daylighting with artificial sources. The central staircase is lit at night from a large pendant chandelier, formed from glass mosque lights specially designed by the architect Neville Conder.

Plans at each level to show the accommodation within (Casson Conder).

Architect: Casson Conder; Lighting Design: DPA Lighting Consultants

Interior of the social room by day (Photographer Derek Phillips).

The council chamber at night (Photographer Derek Phillips).

Exterior view (Photographer Derek Phillips).

The screens to the prayer hall (Photographer Derek Phillips).

The central chandelier designed by the architect (Photographer Derek Phillips).

Case study 59 Royal College of Physicians, Regents Park, London

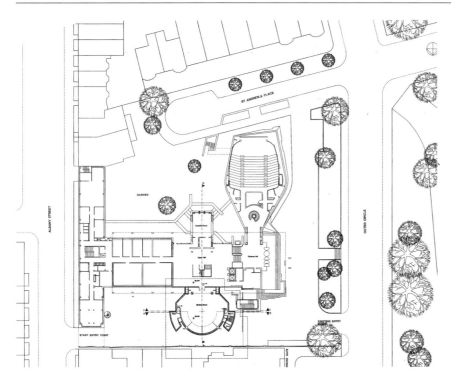

Plan of the extended building (Sir Denys Lasdun).

West elevation

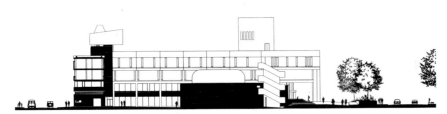

North elevation

Elevation to Regents Park (Sir Denys Lasdun).

The Royal College of Physicians, designed by the architect Sir Denys Lasdun and completed in 1964, replaced the original 1825 building by Robert Smirke in Pall Mall. It is widely recognized as one of the finest 1960s' buildings in London, and despite its cubist influences (or possibly because of them) fits comfortably into the Nash Terraces of its location in Regents Park.

Designed as the headquarters of one of the country's great institutions, the brief to the architect provided a formidable list of accommodation, to include a main circulation space giving access to the library, dining room, lecture theatre and the censor's room, based on the traditional interior to be transferred from the original building. In 1996 Sir Denys added a further lecture theatre and council chamber, the original design of the building being sufficiently robust to accept these changes.

It can be seen from the illustrations that the central circulation or staircase hall that provides access to the two higher floors is flooded with daylight from large side windows which continue upwards to light the hall at high level. Although artificial light from recessed filament fittings is provided at high ceiling level, this is unnecessary except at night or in particularly overcast weather.

The censor's room with its dark wood-panelling derives its daylight from corner slit windows, which give gentle daylight to its classical interior; whilst the lighting of the library is from rectilinear roof lights integrated with artificial light fittings using the compact fluorescent lamp.

The new council chamber, circular in plan, has a concealed roof light over half of the space which floods the rear wall with daylight, the central area with its raised timber roof being uplit to give the necessary artificial light at night. All the light fittings in the chamber use tungsten halogen lamps which allow for dimming and have a control system to give different atmospheres.

The original Woolfson theatre, where no natural light is admitted, uses a series of dropped tiers in the ceiling to conceal fluorescent compact lamps, lighting forward, and controlled by dimmer to

Architect: Sir Denys Lasdun

produce the level of light appropriate to the occasion.

The daylighting of this building would have been seen as superb no matter when it was designed, and certainly the artificial lighting of the original building was state-of-the-art when it was implemented in the 1960s.

The new meeting room (Photographer Niall Clutton).

Entrance view up staircase (Photographer Derek Phillips).

Top floor of the hall (Photographer Derek Phillips).

The library (Photographer Derek Phillips).

Glossary

A simplified explanation of references used in the text divided into eight headings. Use the index for page references.

SEEING/PERCEPTION

Adaptation   The human eye can adapt to widely differing levels of light, but not at the same time. When entering a darkened space from a brightly-lit space the eye needs time to adapt to the general lighting conditions and this is known as 'adaptation'.

Clarity   Clearness, being unambiguous.

Contrast   The visual difference between the colour or brightness of two surfaces when seen together. Too high a contrast can be the cause of glare.

Modelling   The three dimensional appearance of an object, surface or space as influenced by light. Good modelling aids perception.

Perception   Receiving impressions of one's environment primarily through vision, but also using other senses, providing a totality of experience.

Quality   A degree of excellence. The 'quality' of a lighting design project derives from a series of different elements, the most important of which is 'unity,' but which also includes aspects such as modelling, variety, colour and clarity.

Unity   The quality or impression of being a single entity or whole. This can be applied equally to a small or large complex. The word 'holism' is often used in its place.

Variety   The quality of change over time in brightness, contrast and appearance of a space, or series of spaces.

Virtual image   An image of a subject or lit space formed in a computer which can be used to provide a visual impression of the lighting design in order to explain a proposal.

Visual acuity   A measure of the eye's ability to discern detail.

Visual task/task light   The visual element of doing a job of work, and any local or concentrated light fitting placed to improve visibility.

LIGHT SOURCE/DAYLIGHT

Bilateral daylight   Daylight from both sides of a building.

Daylight   The light received from the sun and the sky, which varies throughout the day, as modified by the seasons and the weather.

Daylight factor (DF)   The ratio of the light received at a point within a building, expressed as a percentage of the light available outside. Since daylight varies continually, the amount of light from a given DF is not a finite figure, but gives a good indication of the level of daylight available.

Daylight linking   Controls which vary the level of artificial light inside a building in relation to the available daylight.

No-sky line   The demarcation line within a building where, due to external obstruction and window configuration, no view of the sky is visible.

Obstruction/view   The diminution of available light and view by other buildings at a distance. View is an important environmental aspect of daylighting, which may be impaired by obstruction, but can sometimes be overcome by attention to orientation.

Orientation   How a building relates to the points of the compass. In the northern hemisphere care must be taken with southern exposure.

Shading/*brise soleil*   The means adopted to prevent the deleterious effects of solar gain from southern exposures. These may be external, structural louvres attached to the face of the building or different forms of helioscreen blind.

Sky glare   The unacceptable contrast between the view of the sky outside and the interior surfaces.

Skylight   The light received from the whole vault of the sky as modified by the weather and time of day, ignoring sunlight.

Solar gain   Heat derived from the sun; whilst generally therapeutic, it may require control using a blind, louvre or solar glass.

Solar glass   Glass designed to reflect a percentage of direct heat (infra red) from the sun.

Sunlight   The light received directly from the sun, as opposed to that derived from the sky.

Sunpath   The sun's orbit. As the earth travels around the sun, variations occur both throughout the day and the seasons; these changes in position are known as the sunpath. This can be accurately predicted.

Window   'Wind-eyes' in their many forms provide daylight to an interior.

LIGHT SOURCES OTHER THAN DAYLIGHT/ARTIFICIAL

Arc light   The first form of electric light derived by passing an electric current between two electrodes. Developed by Sir Humphry Davy in 1809.

Candles   Candles are made by moulding wax or other types of flammable material around a wick, which sustains a flame to give light. Modern candles are clean, do not gutter and provide light of a particular quality suitable for social occasions. There have been many light sources attempting to imitate the quality of candlelight; most fail completely, whilst one or two later versions have achieved some success.

Electric light   With the arc lamp and the development of the incandescent lamp in the nineteenth century by Edison and Swan the foundations were laid for all modern forms of light derived from electricity.

Electric light sources   These lamps are described in detail in Chapter 5 and are listed here:

INCANDESCENT SOURCES
tungsten filament
tungsten halogen
low-voltage tungsten halogen

DISCHARGE SOURCES
cold cathode (fluorescent)
mercury fluorescent (high- and low-pressure)
low-pressure sodium
high-pressure sodium
high-pressure mercury
metal halide (including ceramic arc)

FLUORESCENT LAMPS
halophosphor tubular fluorescent
triphosphor tubular fluorescent
compact fluorescent
induction lamps

Fibre optics (remote source)   At its simplest this is the transfer of light from a light source placed in one position to light emitted in another using glass fibre or polymer strands.

Fluorescent phosphors   The internal coatings on surfaces of mercury discharge lamps that produce visible light when excited by the ultraviolet rays emitted by the discharge. The phosphors determine the colour of the visible light.

Gaslight   The light derived from burning coal gas, developed in the late eighteenth century and widely used during the nineteenth century both for domestic and industrial use.

Oil lamps   These, together with firelight, are the earliest forms of artificial light source, the oil being derived from animals, birds or fish. Hollowed-out stone dishes and later clay pots were used with some form of wick. Oil lamps survived until the nineteenth century with the development of the Argand lamp.

LIGHTING TERMINOLOGY

Angle of separation   The angle between the line of sight and the light fitting. The smaller the angle the more likely it is that the light will be glaring.

Brightness   The subjective appearance of a lit surface dependent upon the luminance of the surface and a person's adaptation.

Bulk lamp replacement   The replacement en masse of the lamps in a lighting system when it is calculated that a percentage of the lamps will fail and that the light output of the system will fall below the design level.

Colour   We accept that we only see true colour under daylight, despite the fact that the light varies considerably throughout the day. All artificial sources distort colour in one way or another.

Colour rendering   A comparison between the colour appearance of a surface under natural light and under an artificial source.

Efficiency/efficacy   The ratio of the light output from the lamp to the energy consumed in lumens/watt.

Flicker   The rapid variation in light from discharge sources due to the 50 HZ mains supply, which can cause unpleasant sensations. With the development of high-frequency gear the problem is overcome.

Glare/reflected glare   The most important negative aspect of quality. There are two types of glare: discomfort and disability. Both types are the result of too great a contrast. Glare may result from both daylighting or artificial lighting, either directly or by reflection and must be avoided at the design stage.

Illumination level   The amount of light falling on a surface expressed in engineering terms as lumens per square metre (or lux) and known as illuminance.

Intensity   The power of a light source to emit light in a given direction.

Light fitting/luminaire   The housing for the light source which is used to distribute the light. Whilst the technical word is luminaire, the more descriptive term of light fitting is still widely used. The housing provides the support, electrical connection and suitable optical control for the light source.

Luminance   Light emitted or reflected from a surface in a particular direction; the result of the illumination level and the reflectance.

Lux   The measure of illumination level (illuminance) in lumens per square metre. The Foot Candle is used as a measure in the United States, which is 1 lumen per square foot, or 10.76 lux.

Maintenance factor   The factor applied to the initial illumination level to take account of dirt accumulation and fall-off in light output from the lamp when calculating the level of useful light.

Reflectance   The ratio of light reflected from a surface to the light falling upon it as affected by the lightness or darkness of the surface.

Reflection factor   The ratio of the light reflected from a surface to the light falling upon it. The surface whether shiny or matt will affect the nature of the reflected light.

Scalloping   The effect gained from placing a row of light fittings too close to a wall. Where intentional, this effect can be pleasing, but more generally it becomes an unwanted intrusion on the space.

Sparkle   Refers to rapid changes to light over time and is most readily applied to the flicker of candlelight or firelight. Sparkle can also be created from reflected or refracted light from small facets, such as those of a glass chandelier.

LIGHTING METHODS

Ceiling-/wall-mounted   Light fittings supported directly from the ceiling or wall.

Concealed   Lighting concealed in the ceiling or wall configuration, to provide light on to adjacent surfaces.

Decorative lighting   Lighting designed to be seen and enjoyed for its own sake, such as a crystal chandelier. Alternatively it may be light directed on to objects to achieve a decorative purpose.

Downlight   A light fitting giving its main light downwards. The lights are generally recessed and include both wide beam and narrow angles.

Emergency lighting   The lighting system designed to operate in the event of a power failure to facilitate the evacuation of a building or the continuation of essential services. Various methods are adopted to ensure a suitable source of power.

Floodlighting   Generally refers to the exterior lighting of a building at night by lights with controlled beams placed at a distance.

Functional lighting   Lighting planned to provide light to satisfy the practical needs of a space.

General diffusing   Light fittings giving all-round light.

Indirect   Lighting provided indirectly, reflected from ceiling or wall.

Local light/task light   A light fitting designed to light a specific task, generally at the individual's control.

Louvres/baffles   Means of shielding the light from a fitting or from daylight to eliminate glare. They can be fixed or movable.

Portable   Light fittings such as table and floor standards designed to provide local light. Portable uplights are a useful addition.

Raising and lowering gear   The apparatus applied to heavy light fittings in tall spaces, to allow them to be lowered for lamp change and maintenance.

Spotlight   Light fittings designed to throw light in beams of varying width and intensity.

Suspended   The pendant method of hanging a light fitting from the ceiling or roof.

Torchère   Originally a decorative, free-standing candle holder but now sometimes a term applied to modern wall brackets.

Track-mounted   Light fittings both supported and energized from the numerous track systems available, giving the lighting flexibility.

Uplight   A light fitting directing its light up to the ceiling to provide indirect light.

Wall washing   Lighting where a wall is designed to be lit evenly; several methods can be adopted to achieve this, some more successful than others.

ENERGY AND CONTROLS

BEMS Building Energy Management Systems. A means of computer control of lighting systems within a building.

Control gear Discharge sources require control gear to operate. This includes, amongst others, starters, igniters, transformers, capacitors, ballasts and chokes. Incandescent lamps require no gear, giving low initial cost and making dimming simple.

Digital multiplex controller A sophisticated electronic controller used to monitor and vary circuits in a lighting system, such as might be used in a theatre.

Dimming Dimming controls are exactly what the name implies: devices to reduce the intensity of a light source. All filament sources, both mains and low-voltage, can be controlled by simple dimmers.

Intelligent luminaires Light fittings with inbuilt sensors programmed to vary the light intensity and generally related to the amount of available daylight or occupancy.

Noise attenuation Noise reduction.

Passive building A building that by its configuration eliminates the need for mechanical ventilation and reduces the need for daytime electric lighting.

Photocell Measures illuminance at any position. When placed externally the photocell allows internal light control systems to react to changes in the weather, an element of daylight linking.

Photovoltaics A developing technology where external panels on the southern exposure of a building are designed to convert solar energy into useful electricity.

Scene set More complicated electronic controls that use a microprocessor to create different room appearances at the touch of a button. A number of 'scenes' can be preset and subsequently changed automatically.

Stack effect The way in which hot air will rise in a chimney.

Wind turbine A windmill designed to generate electricity.

ARCHITECTURE

Atrium The courtyard entrance to a roman house, which has an opening in the centre through which rainwater was collected. This opening also provided light to the courtyard and surrounding rooms. The word is now used for multi-storey spaces that are daylit from overhead glazed roofs.

Barrel vault A continuous structural vault of semi-circular section, used from the time of Roman architecture to the present day. Nowadays it is formed of reinforced concrete.

Ceiling coffer A form of concrete roof construction, where, to add strength without increased weight square holes or coffers are added leaving a waffle shape into which services can be placed.

Clerestory (also clear-storey, and pronounced this way) The upper storey with windows above the side aisle roofs that give a high level daylight, particularly in a church.

Conservation The protection of works of art against the deleterious effects of the environment. The control of light levels (particularly ultra-violet) is a major component of conservation.

Dimensional co-ordination The way different building materials are planned to fit together.

Folded plate ceiling A modern development of shell concrete construction.

Glass brick The development of bricks made from glass in the 1930s allowed architects to design structural see-through walls. The Maison Verre in Paris is a well known modern example. Although not widely used today they remain a useful architect's tool.

Lighting gantry A light-weight bridge independent of the main structure of a building that provides support and electric power to light fittings.

Roof monitors/laylights The roof construction in which daylight is permitted to enter a space from overhead. In the case of factories these would be designed to control the entry of sunlight.

Roof truss A development of the beam supports to a roof, that allows openwork lattice to accept services.

Scale Scale is a matter of proportion: the larger the scale, the less human the building will appear. It is sometimes difficult to judge the size of a particular building or interior until a person is added to give it scale.

Shell concrete A thin skin of reinforced concrete, formed in a curve to span the roofs of large areas.

Sprinkler system A method of fire control using a system of water pipes that are designed to deluge water to douse a fire, when design temperatures are exceeded.

Undercroft A term in mediaeval architecture for the lower level vaulting of a cathedral above which the main edifice is built.

Bibliography

BOOKS

Fletcher, Sir Bannister *A History Of Architecture* (20th edition), Architectural Press, 1996

Frampton, Kenneth, *Modern Architecture, A Critical History*, Thames and Hudson, 1980 (third edition 1997)

Gardner, Carl and Hannaford, Barry, *Lighting Design*, Design Council, 1993

Hopkinson, R. G. and Kay J. D., *Lighting of Buildings*, Faber & Faber, 1972

Jodidio, Philip *Tadao Ando*, Taschen, 1997

Tregenza, Peter and Loe, David, *The Design of Lighting*, E & FN Spon, 1998

Loe, David and Mansfield K. P., *Daylighting Design in Architecture*, BRECSU, Building Research Establishment, 1997

Lloyd Jones, David, *Architecture and the Environment*, Laurence King, 1998

Phillips, Derek, *Lighting in Architectural Design*, McGraw Hill, 1964

Phillips, Derek, *Lighting Historic Buildings*, Architectural Press, 1997 (Co-publisher, McGraw Hill)

Rasmussen, Steen Eiler, *Experiencing Architecture*, MIT Press, 1959

Steffy, Gary, *Architectural Lighting Design*, Van Nostrand Reinhold, 1990

LIGHTING PUBLICATIONS

CIBSE, *Code for Interior Lighting*, 1994

CIBSE, *Daylighting and Window Design*. Lighting Guide LG10, 1999

CIBSE, *Lighting for Museums and Art Galleries*, Lighting Guide, 1994

DOE/BRE/RIBA/CIBSE Bell, James and Burt, William, *Designing Buildings for Daylight*, 1995

European Commission, Directorate-General for Energy (Thermie), *Daylighting in Buildings*, 1994

European Commission, Directorate-General for Energy (Thermie), *Energy Efficient Lighting in Buildings: Offices/Industrial/Schools*

Lighting Industry Federation (LIF), Lamp Guide, 1998

RIBA/Thorn, *Electric Lighting for Buildings*, Lynes & Bedocs (Professional Studies in British Architectural Practice)

JOURNALS

The International Lighting Review, Philips, Eindhoven

Light and Lighting, magazine of the CIBSE lighting division
Lighting Journal, Institute of Lighting Engineers (ILE)
Lighting Equipment News, EMAP
Light, ETP Ltd

PAPERS/ARTICLES

Bell, James 'Development and Practice in the Daylighting of Buildings'. Trans. Illum. Eng. Soc. 5, No. 4, 1973

Littlefair, Paul 'Innovative Daylighting'. Review of systems and Evaluation Methods. LR and T/CIBSE 22, No. 1, 1990

Loe, David and Rowlands, E 'The Art and Science of Lighting: A strategy for Lighting Design'. LR and T/CIBSE 28, No.4, 1996

Parry, Malcolm 'Lighting in the Architecture of Sir John Soane'. CIBSE LR & T. 29, 3 965–104, 1997

Phillips, Derek 'Space Time and Light in Architecture'. Presidential Address to the IES 1974. Trans. Illum. Eng. Soc. 7, No. 1, 1975

Phillips, Derek 'Architecture - Day and Night'. CIBSE National Lighting Conference, 1992

Phillips, Derek 'Lighting and Interior Architecture'. Lux Europa, 1993

Ribeiro JS, Allen WA, and M De Amorim 'Lighting of the Calouste Gulbenkian Museum'. LR & T, Trans. Illum. Eng. Soc. 3, No. .2, 1971

Index of architects and designers

Index of lighting designers

Index of subjects

Entries in capitals refer to Glossary headings. Entries in bold refer to Case Studies.